Optical Microscopy of Materials

Optical Microscopy of Materials

R. HAYNES B.Met., Ph.D., C.Eng., F.I.M., M.Inst.P.

Department of Materials Engineering and Design
University of Technology
Loughborough

International Textbook Company

Published by International Textbook Company
a member of the Blackie Group
Bishopbriggs, Glasgow G64 2NZ, and
Furnival House, 14–18 High Holborn, London WCIV 6BX

First published 1984

British Library Cataloguing in Publication Data

Haynes, R.
Optical microscopy of materials.
1. Materials—Microscopy
I. Title
620.1'1'028 TA418.62

ISBN 0-7002-0287-0
ISBN 0-7002-0286-2 Pbk

Printed in Great Britain by McCorquodale (Scotland) Ltd.

Contents

Preface

Since Sorby published his observations on the structures of steels in 1863, the optical microscope has become one of the most widely used and versatile instruments for examining the structures of engineering materials. Moreover, to examine the diverse range of materials encountered, it must be used in both the reflected-light and transmitted-light forms, and with polarized light. It is complementary to, but not superseded by, the wide range of electron-optical instruments that are now used.

Despite its extensive use, it has been described as the most misused, abused, and misunderstood of scientific instruments, for it will produce an image of a sort no matter how badly it is used. To use it effectively, even in its simplest applications, a knowledge of the simple theory of the microscope is necessary, for the theory shows and explains how it should be used. Thus my aim has been to give a simple and, where possible, quantitative account of both the theory and the use of the microscope, including the various special techniques for which it can be used.

But, no matter how effectively the microscope is used, if the specimen is inadequately prepared the results of examination will be of doubtful value. Therefore, I have included a chapter on specimen preparation that sets out the range of methods available, and describes the methods that are most widely used, in sufficient detail to enable the reader to carry them out. In this chapter, I have departed somewhat from my intention to avoid wherever possible the use of references in the text in order to draw the reader's attention to useful sources of practical information, such as compositions of polishing solutions and etchants, that could not be included in the text because of limitations of space.

The book developed from lectures and laboratory work that I have given to honours degree students in materials engineering and metallurgical engineering for many years. However, I hope that it will be equally useful to undergraduates in other materials disciplines, technicians and postgraduate workers in the engineering materials field. Indeed, on many occasions research students have remarked on the need for a simple brief introduction to optical microscopy, such as I have tried to produce. In preparing my lectures and writing the book I have drawn not only upon my own experience, but upon the extensive literature of optical microscopy, and I am indebted to the many authors whose work I have consulted.

R.H.

1 The mechanical construction of the microscope

There are two types of microscope, simple and compound. A simple microscope is a magnifying glass and usually is designed to produce an image that is up to several times larger than the object. Such glasses are useful for visual examination of macroscopic features, e.g. of fractures. However, to observe and resolve finer detail a compound microscope must be used.

Essentially the compound microscope consists of two compound lenses, the objective and the eyepiece. The former produces a magnified real image of the object, that is further magnitude by the latter. However, since most objects are not luminous, an illuminating system needs to be included in the apparatus and is often incorporated in the microscope unit. The type of illuminating system used depends on whether the object is to be viewed by transmitted or reflected light.

Compound microscopes can be divided into two groups; those that are intended mainly for direct visual examination of the image, called bench microscopes, and those that are intended for both direct visual examination and photographic recording of the image, called projection microscopes. The latter, usually much more sophisticated than the former, are designed to enable a wide variety of microscopical techniques to be used, and often can be used with both transmitted and reflected light.

1.1 The bench microscope

In its simplest form the lenses of a compound microscope are mounted at the opposite ends of the microscope body tube, which is painted matt black inside to minimize internal reflections that would contribute to glare. On modern microscopes the tube is usually of fixed length, the mechanical tube length, because objectives are designed to work best at a fixed mechanical

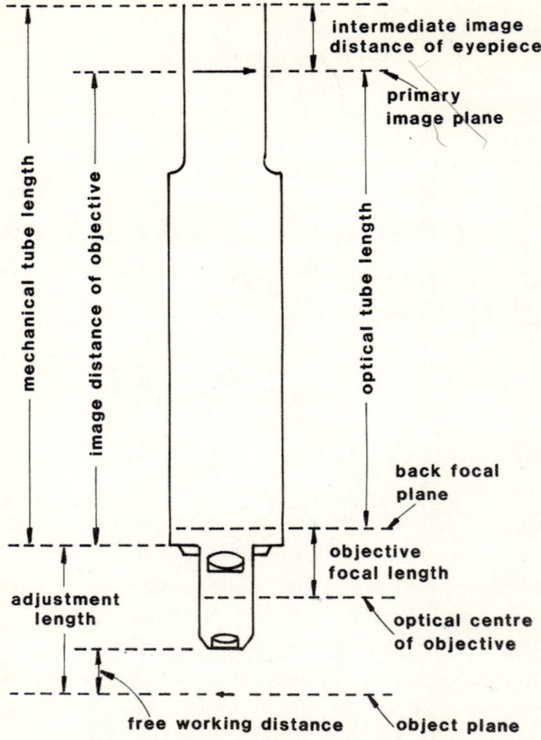

Figure 1.1 Relationships between optical and mechanical distances in the compound microscope.

tube length.* Most microscope makers have standardized on a mechanical tube length of 160 mm, but Leitz use 170 mm. However, a fairly recent innovation is the use of objectives that are corrected for infinite tube length. These objectives produce real intermediate images in the tube only in conjunction with a tube lens built into the microscope tube. The tube lens may alter the magnification produced by the objective by a small factor, commonly × 0.8, × 1.0, or × 1.25. Figure 1.1 summarizes the relationships between optical and mechanical distances in the microscope. Note that the mechanical tube length usually differs markedly from the optical tube length, and should not be substituted for the latter in optical calculations.

The microscope is mounted on a heavy, well-balanced, rigid stand to minimize vibrations that would cause an unsteady image (Fig. 1.2). The stand comprises a foot, and a limb that carries the microscope, the stage,

* Older microscopes often are equipped with draw tubes that enable the mechanical tube length to be adjusted to a value appropriate to the objective.

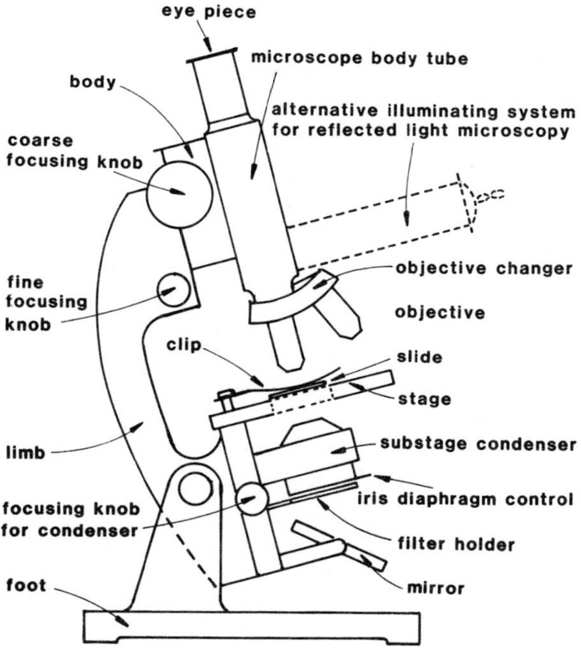

Figure 1.2 Diagram showing the components of a bench microscope. Microscopes are equipped with an illuminating system for either transmitted light (substage attachments shown in full line), or reflected light (shown in broken line), usually not both.

and, if required, the substage. The limb is hinged to the foot, which traditionally is either of horseshoe shape or of tripod form to permit the instrument to stand firmly without rocking. In transmitted-light microscopes the body contains the focusing mechanism, but in reflected-light microscopes, usually called metallurgical microscopes, the mechanism more commonly is fitted to the stage. Usually, it consists of two parts, a rack and pinion mechanism for coarse focusing and a lever or cam mechanism for fine focusing. Both of these should be free from sideways movement and backlash, which make accurate focusing difficult. Two large milled knobs, one on each side of the instrument, operate the coarse focusing, while two smaller knobs operate the fine focusing.

Usually a revolving nosepiece, or objective changer, that carries three to five parfocal objectives arranged in order of magnification, is attached to the lower end of the body tube. A click stop positions each objective correctly beneath the tube. The eyepiece slides into the upper end of the body tube and is supported by its flange.

The stage is a rigid platform, attached to the lower part of the limb, and this must be perpendicular to the optic axis of the microscope, otherwise the focus varies from edge to edge of the field. For illumination by transmitted light there is a hole in the centre of the stage, but this is unnecessary for illumination by reflected light. It is provided with either clips to hold the microscope slide, or a mechanical stage that holds the slide and enables it to be moved smoothly and accurately in two directions at right angles by means of rack and pinion mechanisms.

For examination by transmitted light the microscope is equipped with a substage that carries a substage condenser, filter holders that can be swung out of the optical path, and a hinged mirror, with one side plane and the other concave, to reflect light into the condenser (Fig. 1.2). The condenser is fitted with adjustment screws that allow it to be centred on the optic axis of the microscope, and with a rack and pinion mechanism that permits it to be focused on the object plane.

Alternatively, for examination by reflected light the illuminating system commonly is incorporated in a side arm attached at right angles to the microscope tube (Fig. 1.2, broken lines). The side arm carries the light source, a condenser lens or lenses, and often two iris diaphragms to control the illumination of the object. Light from the side arm enters the microscope tube and is partially reflected through the objective onto the specimen by a thin glass slip inclined at 45° to the optic axis of the microscope. The light reflected from the specimen returns through the objective and is partially transmitted to the eyepiece by the glass slip.

In many modern bench microscopes the limb and foot are integral, and the body, no longer a simple tube, is rigidly fixed to the stand. A prism in the body reflects the light into the eyepiece, that is at an angle of about 135° to

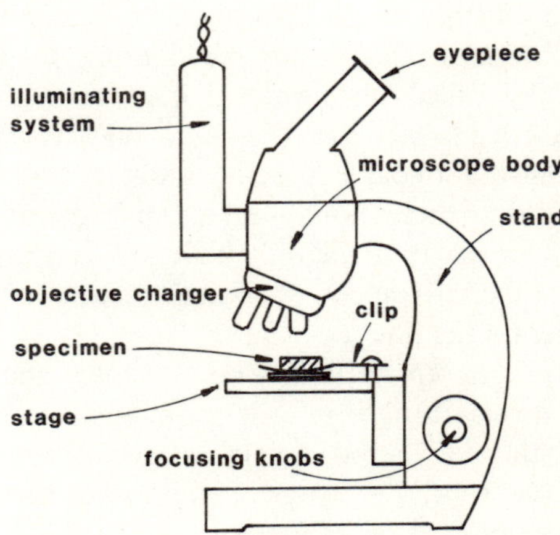

Figure 1.3 A modern design of bench metallurgical microscope.

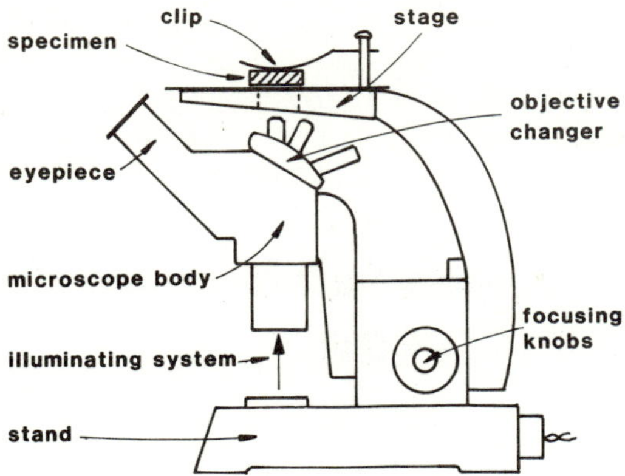

Figure 1.4 Diagram of an inverted metallurgical microscope.

the lower part of the body, which is vertical, eliminating the need to tilt the limb (Fig. 1.3). However, in order to maintain the appropriate mechanical tube length there is included in the optical system a tube lens that may make a small contribution to the overall magnification. Focusing is achieved by moving the stage. Sometimes a beam splitter is incorporated in the body so that the image can be viewed by both eyes through a matched pair of eyepieces.

For metallographic work an inverted bench microscope is sometimes used (Fig. 1.4). In this case the body of the microscope is arranged so that the objective lens points upwards and a prism in the body reflects the light from the specimen into an eyepiece inclined at about 45° to the optic axis of the microscope. The stage is above the objective lens and focusing is achieved by moving the stage. The light source is built into the foot of the microscope. This design of microscope enables larger, heavier specimens to be examined than can be accommodated on the stage of a microscope of conventional design.

2 Image formation in the microscope

In order to understand the formation of the image in the microscope we must consider it from the points of view both of geometrical optics and of the wave theory of light. The former yields information about the magnifications that can be attained and the types and positions of the image formed, while the latter yields information about the detail that can be resolved by the microscope objective and the quality of the image.

2.1 Refractive index, refraction and reflection

When light travels through vacuum its velocity is $3.00 \times 10^8 \text{ ms}^{-1}$, but when it traverses other media its velocity is reduced. The ratio of its velocity in vacuum to that in another medium is called the *refractive index*, *n*, of the other medium. Air and gases have refractive indices which are virtually indistinguishable from that of vacuum, i.e. 1.00. Common transparent liquids and solids have values of refractive index between 1.3 and 1.7, although some substances, including some fairly recently developed optical glasses, have higher values, sometimes exceeding 2.0. For a given material the refractive index varies with the wavelength of the light; this is called *dispersion*.

When the speed of light waves is reduced their frequency is unaltered but their wavelength is shortened. Consequently, when a ray of light passes from one isotropic medium to another it is refracted, i.e. bent, through an angle which is determined by Snell's Law:*

$$\frac{\sin i}{\sin r} = \frac{v_1}{v_2} = \frac{n_2}{n_1} \tag{2.1}$$

* Snell's Law is readily deduced from wave theory as follows. Let the velocity of light in vacuum be c and in the other medium be v. Consider two parallel rays $A'A$ and $B'B$ with a common wavefront AB impinging on a planar interface between vacuum and the medium at an angle of incidence i, and leaving the interface at an angle of refraction r (Fig. 2.2). Let the

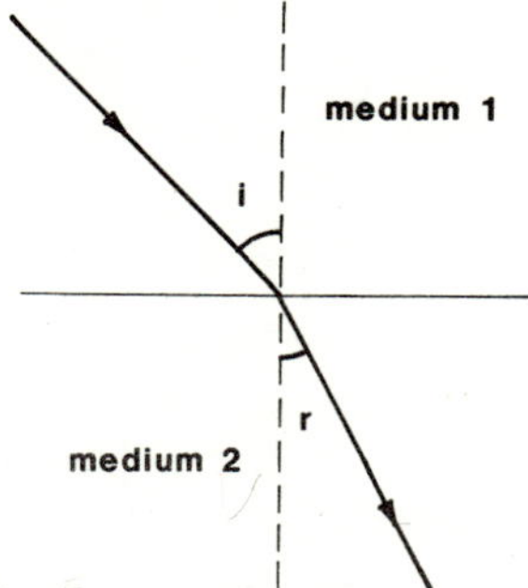

Figure 2.1 Refraction of light at the interface between two media of different refractive index.

where i and r are the angles that the incident and refracted rays make with the normal to the surface between the two media (Fig. 2.1); v_1 and v_2 are the velocities in, and n_1 and n_2 the refractive indices of, media 1 and 2 respectively. The incident and refracted rays and the normal to the surface are coplanar. If the first medium is vacuum or gas, equation (2.1) becomes

$$\frac{\sin i}{\sin r} = n \qquad (2.1a)$$

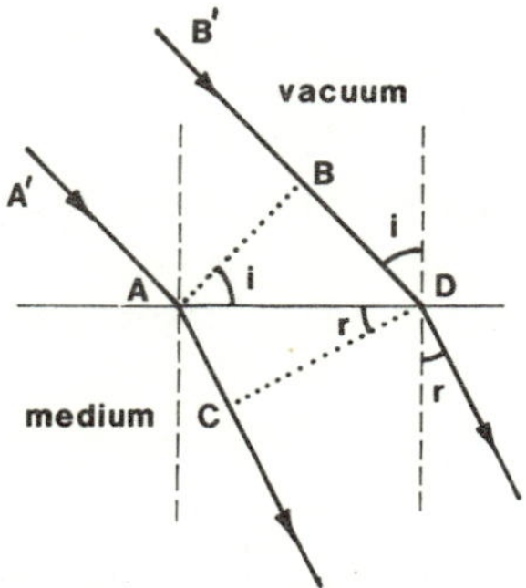

Figure 2.2 Derivation of Snell's Law from wave theory.

time for the wavefront to travel from AB to CD be t. Then

$$BD = ct = AD \sin i$$

$$\text{and } AC = vt = AD \sin r$$

Eliminating AD we obtain

$$\frac{\sin i}{\sin r} = \frac{c}{v} = n \qquad (2.1a)$$

In general, because $v = c/n$, when the refractive indices of the two media are n_1 and n_2, we have

$$\frac{\sin i}{\sin r} = \frac{n_2}{n_1} \qquad (2.1)$$

Similarly, when light is reflected at a surface it obeys another simple law. This is that the incident and reflected rays make equal angles with the normal to the reflecting surface and are coplanar with it.

2.2 Geometrical optics of the compound microscope

In geometrical optics we make use only of the simple idea that light travels in straight lines unless it is refracted or reflected. Image formation in the compound microscope is most easily understood by considering the microscope to be composed of two thin convex lenses (Fig. 2.3), one (OO) acting as the objective and the other (EE) as the eyepiece. Using the rules that rays passing through the optical centre of a lens pass undeviated, while those parallel to the optic axis of the lens must pass through the conjugate focal point of the lens, the ray diagram for the microscope can be drawn (Fig. 2.3). From the thin lens equation

$$\frac{1}{f}=\frac{1}{u}+\frac{1}{v} \tag{2.2}$$

where f = the focal length of the lens, u = the object distance, and v = the image distance; and using the real-is-positive convention, we find that real magnified images are only produced when the object is between one and two focal lengths, f_O, in front of the objective. Moreover, high magnifications are possible only when the focal length of the objective is small. The objective produces a real inverted magnified image of the object AB at $A'B'$, which acts as the object for the eyepiece. It is arranged that $A'B'$ lies just within the object focal distance, f_e, of the eyepiece. Consequently, when placed at the exit pupil of the eyepiece, the eye sees a virtual image at $A''B''$. The virtual image $A''B''$ is magnified but not reinverted and its exact

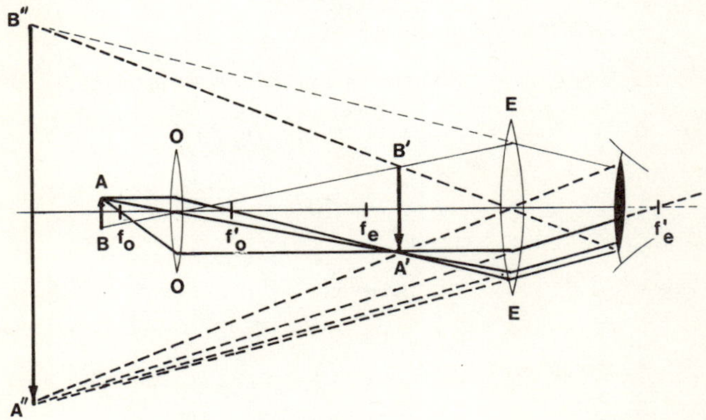

Figure 2.3 Ray diagram illustrating image formation in the compound microscope.

position depends on how the microscope is focused. Usually we assume that it is focused within the image at the nearest distance of distinct vision, which is about 250 mm.

The magnification M produced by a lens is given by the equation

$$M = \frac{v}{u} \tag{2.3}$$

Combining this with equation (2.2) yields for a real image

$$M = \left(\frac{v}{f} - 1\right) \tag{2.4a}$$

and for a virtual image

$$M = \left(\frac{v}{f} + 1\right) \tag{2.4b}$$

Thus for the objective we have

$$M_0 = \left(\frac{v_0}{f_0} - 1\right) = \left(\frac{v_0 - f_0}{f_0}\right) \tag{2.5a}$$

where the subscripts O refer to the objective, and since the optical tube length (Fig. 1.1) $t = v_0 - f_0$,

$$M_0 = \frac{t}{f_0} \tag{2.5b}$$

Hence, the magnification produced by the objective is fixed and depends on the focal length of the objective. Similarly, the magnification, M_e, produced by the eyepiece is

$$M_e = \left(\frac{v_e}{f_e} + 1\right) = \left(\frac{v_e + f_e}{f_e}\right) \tag{2.6a}$$

where the subscripts e refer to the eyepiece. Assuming that the image is formed at the nearest distance of distinct vision, D, we can put $D = v_e + f_e$, since the eye is placed at the exit pupil of the eyepiece, thus

$$M_e = \frac{D}{f_e} \tag{2.6b}$$

As in the case of the objective, the magnification produced by the eyepiece is fixed and depends on its focal length. Hence the total magnification, M, produced by the microscope is

$$M = M_0 \times M_e = \frac{tD}{f_0 f_e} \tag{2.7a}$$

However, if a tube lens is incorporated in the microscope the total magnification is

$$M = M_0 \times M_e \times M_t \tag{2.7b}$$

where M_t is the magnification produced by the tube lens. Approximate values of M_0, M_e, and M_t are marked on objectives, eyepieces, and tube lens mounts and their product gives a value for magnification that is acceptable for many purposes, e.g. designating the magnification of a photomicrograph.

If an accurate value of magnification is required, as when measuring the size of features in a microstructure, it must be measured by means of a stage micrometer. This consists of a series of equispaced parallel lines of known spacing ruled either on a polished metal surface for reflected light microscopy, or on a glass surface for transmitted light microscopy. The magnification is found by measuring the spacing of the lines in the magnified image and dividing by the true spacing.

2.3 Coherent and non-coherent light

Geometrical optics enables us to understand the way in which a microscope produces a magnified image, but it tells us nothing about the quality of the image that is formed. To understand this we must consider image formation in terms of the wave theory of light.

Separate waves of the same frequency interfere, either reinforcing or annulling each other depending on their phase, but waves with different frequencies do not interfere and behave as if they were in isolation. Rays which are able to interfere are described as being *coherent*, while those that cannot are described as being *non-coherent*.

The eye is able to distinguish differences in the amplitudes of waves because the intensity of a ray of light is proportional to the square of its amplitude, and to distinguish differences in wavelength because change in wavelength corresponds to change in colour, but it cannot distinguish differences in phase. Therefore, if as in some special microscopy techniques, we wish to make use of differences in phase to distinguish between different constituents in a microstructure, the difference in phase must be translated into differences in amplitude.

Truly coherent light, in which the waves are all exactly in phase, is only obtained from individual atoms and from lasers, in which many emitters are constrained to vibrate in phase. Lasers are very powerful sources of coherent radiation, but to date have found only very limited use in microscopy.

Fortunately, the phases of rays emitted by an incandescent light source can be considered to combine to produce a composite waveform that is unique to the source at any instant, but the waveform will vary with time. Hence, provided that the source subtends a small angle at the point of interest so that it approximates to a point source, each ray from it will have an identical waveform and the beam can be regarded as being coherent: such sources are called *compact* sources. A compact source of monochromatic light radiates in all directions light waves which are all of the same frequency and which are all in phase. Thus at all points on a spherical surface having the source as its centre, the waves will be in phase. Almost all light sources used in microscopy are of this kind, including the widely used tungsten filament lamps, arc lamps, mercury and xenon lamps. Even the traditional light source, the sun, is of this kind because it is so far away that it subtends only a very small angle at the point of interest.

We can consider any point on an object to be a light source emitting waves in all directions, and an object to be made of many such sources. If all the waves leaving a point on the object could recombine, a perfect image of the point would form. In practice this cannot be achieved because many of the waves travel in directions from which they cannot be collected by the lens forming the image and are lost (Fig. 2.4). Thus the perfection of the image depends on the amount of light from the object collected by the optical system, provided that the rays are focused accurately and are in phase.

The phase relationship of the rays is especially important when they are coherent. When two coherent rays combine, the amplitudes of their waves are additive. Hence, since the brightness of the light is proportional to the square of the amplitude of the waves, when two waves of equal amplitude are exactly in phase the brightness of the resultant wave is four times that of either of the original waves. Conversely, when they are exactly out of phase the light is extinguished. On the other hand, non-coherent waves

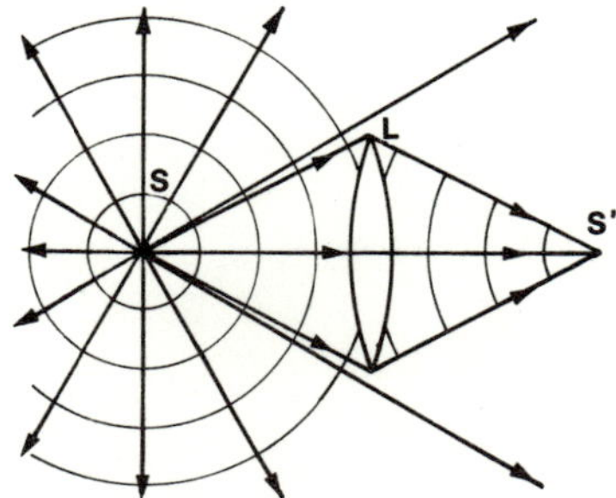

Figure 2.4 Sketch showing why a lens is unable to form a perfect image. Source S emits spherical waves only part of which fall on the lens L, hence the image S' formed by the lens is imperfect.

do not interfere or combine. Thus when two non-coherent waves of equal amplitude propagate, the intensity is simply twice that of either of the original waves.

When, in an image formed by coherent light, the waves fail to remain in phase the quality of the image deteriorates, and the image may be destroyed. Waves remain in phase when their optical paths contain the same number of waves. For example, consider the light passing through a convex lens; the axial ray has the shortest path through air and the longest path through glass, while the marginal rays have the longest path through air and the shortest path through glass. The wavelength in glass is shorter than that in air, thus in principle the lens is designed so that all paths are optically equal. Nevertheless in practice it is sufficient that the converging rays remain in phase within a quarter of a wavelength.

2.4 Diffraction

Diffraction is the most important image-forming mechanism. It occurs when light passes an edge and causes the shadow to have a diffuse edge (Fig. 2.5). The effect is influenced by the wavelength of the light, longer waves being diffracted more than shorter ones. Since microscope images are formed by diffracted rays it follows that small points are more easily resolved by light of short wavelength (blue light) than by light of long wavelength (red light).

Consider a wide slit illuminated with coherent (monochromatic) light (Fig. 2.6). As the light emerges from the slit it spreads sidewise. When the edges of the slit are brought closer together the spread becomes greater and eventually light and dark side bands appear on either side of the direct beam. In a similar manner, when a pinhole is illuminated the direct beam is surrounded by concentric alternate dark and light rings, known as the Airy disc. The direct beam is called the zero-order beam, and successive diffracted beams are described as first-order, second-order and so on. The effect of a single slit can be reinforced by those of other regularly spaced slits with an interslit spacing equal to the width of the slit.

When two similar parallel slits are illuminated by coherent light the

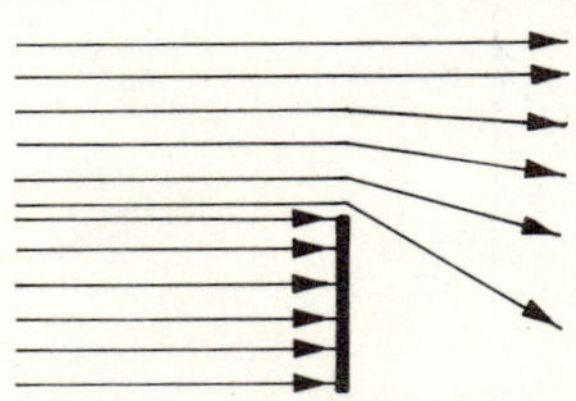

Figure 2.5 Diffraction of light at an edge.

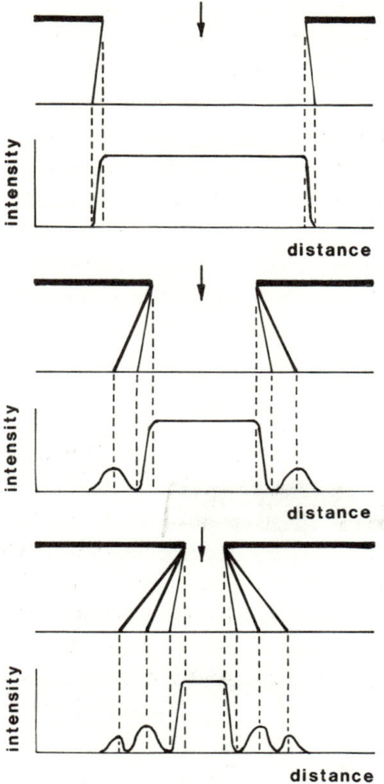

Figure 2.6 Diffraction of light at a slit. As the width of the slit becomes narrower rays are diffracted through larger angles and eventually separate diffracted beams are formed on either side of the slit. As the order of the diffracted beams increases their intensity decreases quickly.

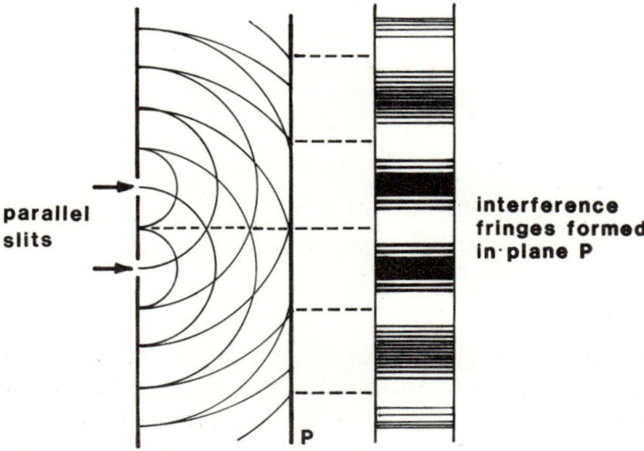

Figure 2.7 Formation of fringes by interference of coherent light from two parallel slits.

diffracted waves interfere (Fig. 2.7) producing a series of alternate light and dark bands parallel to the slits. The centre bands of the striped field is bright and is opposite the midpoint of the slits, i.e. equidistant from both. Dark bands occur where the waves from the two slits are out of phase by half a wavelength. However, when white light is used, the various colours extinguish in order of wavelength, producing on either side of the direct beam a series of Newton's bands which run from blue to red outwards. The angular displacement of successive spectra depends on the spacing of the slits and on the refractive index of the medium in which the spectra are formed; the closer together the slits and the lower the refractive index of the medium the greater the angular displacement.

2.5 Abbe's theory of the microscope

Consider the case of image formation in the transmitted-light microscope. In Fig. 2.8 a transparent object *A* with a fine two-dimensional periodic structure is illuminated by a collimated (parallel) beam of light. The structure diffracts light, as described previously. Thus, from a point P_0 the objective, *O*, collects the direct beam and some of the diffracted beams. Since the object is illuminated with collimated light the direct and diffracted beams are also collimated. Consequently, after passing through the objective all the beams are brought to focus in the back focal plane of the

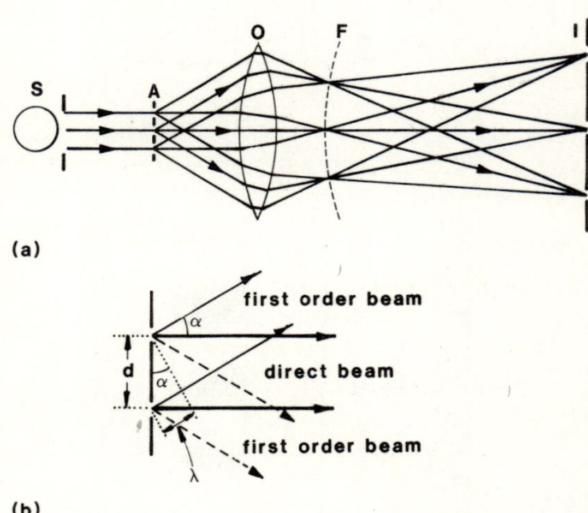

Figure 2.8 Abbe's theory of image formation in the microscope. (*a*) The collimated beam of light from source *S* is diffracted by object *A*. Direct and diffracted beams are collected by the objective *O* and images of *S* are formed in the back focal plane *F* of the objective. These images act as subsidiary sources which emit waves that form an image of *A* in the image plane *I* by interference. (*b*) Criterion for the formation of first-order diffracted beams at the object. $\lambda/d = \sin\alpha$.

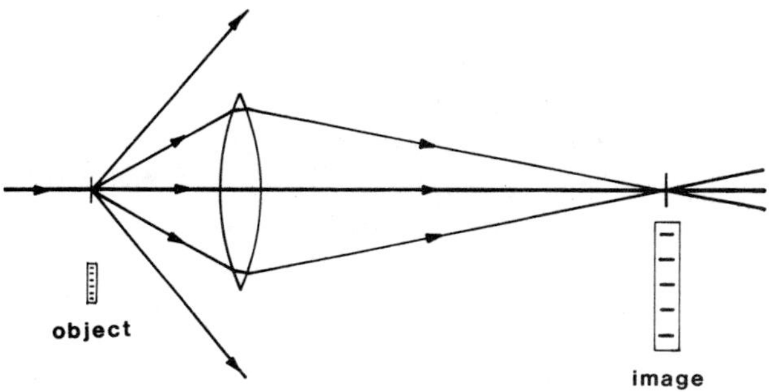

Figure 2.9 Direct and first-order diffracted beams are collected by the objective. Detail in the image is resolved.

objective, F, each beam forming an image of the light source, S. These images of the light source are Huyghenian subsidiary sources, each of which sends out waves that form by destructive interference the image of the point P_0 in the image plane I at point P_i. The subsidiary sources lie on the arc of a circle centred at P_i, so that all the subsidiary waves reach P_i in phase and reinforce each other. Thus, there is produced a bright image of P_0 that is surrounded by alternate dark and light rings formed by interference. The central bright disc is the true image of P_0 but the unavoidable surrounding bright rings reduce its sharpness.

We may draw three main conclusions from the detailed theory. Firstly, at least two beams must be collected by the objective if the detail in the image is to be resolved, because at least two beams are needed for interference to occur (Fig. 2.9); if only the direct beam is collected no detail is shown in the image (Fig. 2.10). Secondly, to form the best possible image all the beams diffracted by the object ought to be collected by the objective and used to form the image. In practice this is impossible but, fortunately, because the

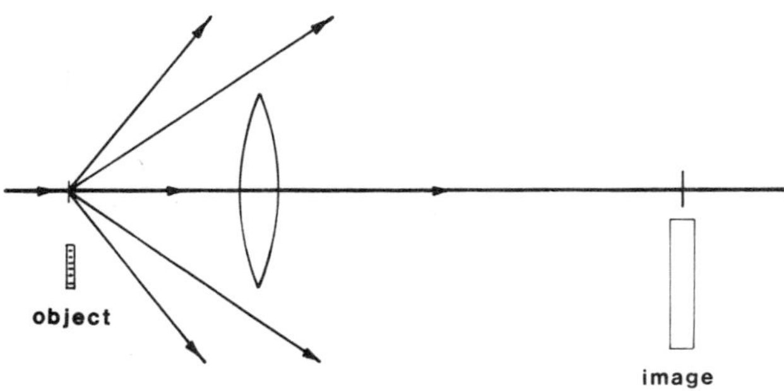

Figure 2.10 Only the direct beam is collected by the objective. Detail in the image is not resolved.

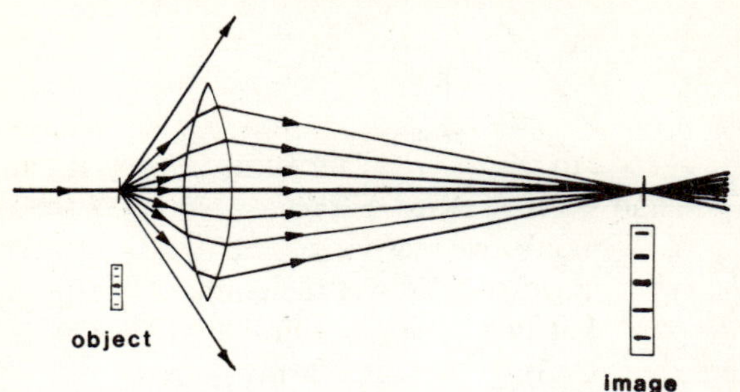

Figure 2.11 The direct and several orders of diffracted beams are collected by the objective. Detail is resolved accurately.

intensities of diffracted beams quickly decrease as their order becomes higher their exclusion does not seriously impair the quality of the image. Thirdly, the sharpness and intensity of the central bright disc depend upon collecting the direct beam and an optimum number of diffracted beams.

In discussing the fringes produced by parallel slits (section 2.4) we noted that the angular displacement of successive spectra was greater the smaller the spacing of the slits. Similarly, we find that the finer the diffracting structure the greater become the angles between the direct beam and the diffracted beams. Consequently, the finer the diffracting structure the wider must be the angle of the cone of acceptance of the objective if the first-order diffracted beams are to be collected. Moreover, the greater the angle of the cone of acceptance, which is measured by the numerical aperture, the more likely are higher-order diffracted beams to be collected by the objective, thus forming a more accurate image (Fig. 2.11).

If the angle between the direct beam and the first-order diffracted beams is so big that the latter cannot be collected by the objective when the objective is illuminated normally (Fig. 2.10), it may be possible to enable the

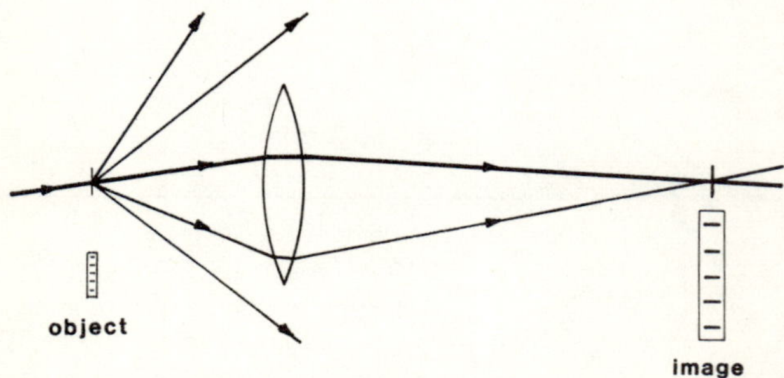

Figure 2.12 By illuminating the object in Fig. 2.10 obliquely the objective collects the direct beam and a diffracted beam. Resolution of detail in the image is restored.

objective to collect the direct beam and a diffracted beam by illuminating the object obliquely (Fig. 2.12). This permits an image to be formed but its quality is poor. As the obliquity of the illuminating beam increases, the ability of the objective to resolve detail increases because the angle between the direct beam and the first-order diffracted beam that the objective can accept increases. Eventually we reach a point at which the direct beam no longer enters the objective, so that only the diffracted beams are collected. However, provided that two or more diffracted beams are collected an image is formed (Fig. 2.13). This technique produces reversed contrast and is described as dark-ground illumination.

The presence of direct and diffracted light beams in the microscope can be demonstrated readily. Using transmitted light, focus the microscope on a thin section and then remove it. Remove the eyepiece and inspect the back of the objective. With the aperture stop diaphragm closed down we see at the centre of the back of the objective a bright disc of light surrounded by an almost completely dark annulus; the bright disc is the direct beam. When we replace the thin section beneath the objective the bright disc remains but the outer parts of the back of the objective are filled with light that has been diffracted by the thin section. Even though the direct beam occupies only a small part of the objective aperture the whole of the aperture is available to the diffracted beams. Opening or closing the diaphragm alters the size of the direct beam but has little effect on the diffracted beams. We can also demonstrate the effect using the metallurgical microscope by first examining a highly polished specimen that diffracts little light, and then a polished and etched specimen.

Furthermore, we can demonstrate the formation of images of the light source in the back focal plane of the objective. Using a low-power objective,

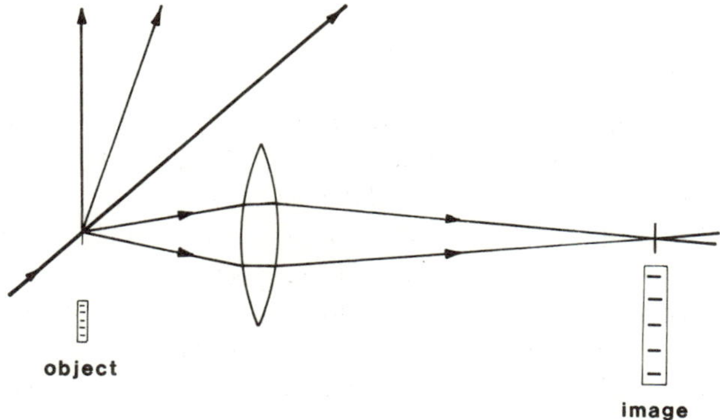

Figure 2.13 The direct beam is so oblique that it no longer enters the objective, but two diffracted beams are collected by the objective. Detail in the image is resolved, but with reversed contrast, i.e. a dark-ground image is formed.

focus the microscope on a stage micrometer, remove the eyepiece and inspect the back of the objective, with the eye over the centre of the objective. When the aperture stop is closed to a small point we see a bright disc, the direct beam, at the centre of a row of fainter coloured discs. The fainter discs are red on one side and blue on the other and are formed by interference between the beams diffracted from the regularly spaced lines on the stage micrometer, which acts as a diffraction grating. The number of discs that we see depends on the spacings of the lines on the stage micrometer and the acceptance angle of the objective. They become more clearly separated the longer becomes the focal length of the objective. When we open the aperture stop the discs grow bigger and overlap, if they did not do so at first, and eventually when the objective is completely filled by the direct beam they cannot be seen. The line of discs is perpendicular to the direction of the rulings and when we rotate the stage micrometer the line of discs also rotates. Moving the stage micrometer without rotating it does not alter the positions of the discs, but they disappear when we move the rulings out of the field of view.

2.6 Subsidiary mechanisms of image formation

While diffraction is the essential mechanism of image formation in the microscope certain other mechanisms may also be involved. These are reflection, absorption, refraction, fluorescence, and polarization. With the exception of polarization, which is the basis of a special branch of microscopy considered in a separate chapter, these are considered below. Moreover, special techniques, involving the shifting of beams in order to produce contrast, are sometimes employed and these too are considered in separate chapters.

2.6.1 *Reflection*

Reflection is an essential mechanism in the formation of images of metallic and other opaque specimens. It occurs at interfaces at which there is a large difference in the refractive indices of the media, as between air and metal. In the metallurgical microscope the specimen may be illuminated either vertically, i.e. normal to the surface of the specimen along the optic axis of the microscope, or very obliquely, producing bright-field and dark-field illumination respectively. In the former, light from the surfaces of grains that are perpendicular to the incident beam is reflected back into the objective, while light from features in the microstructure, such as grain boundaries and tilted areas of the surface, is reflected away from the objective, so that they appear dark. Conversely with dark-field illumination, light reflected from surfaces which are almost perpendicular to the optic axis of the microscope is reflected away from the objective, while light from features in the microstructure is reflected into it.

Reflection also occurs in transparent materials at interfaces between constituents of different refractive index, e.g. grain boundaries. It makes an important contribution to image formation in polycrystalline specimens when dark-ground illumination is used.

2.6.2 *Absorption*

Absorption is important in transmitted-light microscopy because it may help to distinguish between phases. In general, different phases absorb light to differing extents which causes them to differ in brightness in specimens of uniform thickness; however, variations in thickness also produce differences in brightness. Furthermore, in some cases selective absorption of a particular wavelength or wavelengths of light occurs causing the phases to appear coloured. The colours observed are subtractive colours, i.e. white minus the colours absorbed. In biological and polymer microscopy this effect is produced by selective staining of features in a specimen, but usually this course of action is not open to those working with other materials.

When examined in plane polarized light, but not with crossed polars, some non-cubic crystals show selective absorption that varies with vibration direction. This causes the colour of the crystals to change when they are rotated on the microscope stage, e.g. hornblende. The phenomenon is called pleochroism.

In some cases selective absorption plays a part in reflected-light microscopy, thus copper appears a salmon-pink colour. The constituent selectively absorbs particular wavelengths of the light falling on it, causing it to show the subtractive colour. Such absorption often helps to distinguish between and identify constituents, e.g. in aluminium alloys. However, films formed on the surface also may cause constituents to appear coloured, owing to either absorption or interference, as in interference layer microscopy.

2.6.3 *Refraction*

Refraction is used mainly in the examination of thin sections of minerals and in chemical microscopy. The boundaries between colourless transparent constituents are only distinguishable when the constituents differ in refractive index from each other or the medium in which they are immersed. Moreover, the greater the difference in the refractive indices of the constituents the more apparent do the boundaries become. This fact is made use of to determine whether a material has a refractive index smaller or greater than the medium in which it is mounted in the Becke test*, and in the accurate determination of the refractive index of grains of a material by immersing them in liquid media of accurately known refractive indices.

2.6.4 *Fluorescence*

This occurs when a substance absorbs light of one frequency and re-emits light of another, lower frequency and of longer wavelength. It is of interest because it is one of the few cases where an object is self-luminous. In practice, because a fluorescent substance must absorb light before it can re-emit light, specimens are illuminated usually with ultraviolet light using a dark-ground illuminator, so that the beam that excites the fluorescence does not enter the objective. The phenomenon is rarely observed in materials of interest to the materials technologist, but it is sometimes exploited in the biological field where fluorescent stains are used.

* *The Becke Test.*
Using a high-power objective, focus the image of the specimen and then reduce the light intensity to accentuate the Becke line. When we throw the image slightly out of focus a bright rim associated with the boundary between the crystal and the mounting medium appears: this is the Becke line. Slowly move the specimen *away from* the microscope, e.g. by lowering the stage, and observe the movement of the line. It moves from the medium of lower refractive index to that of higher refractive index. The lighting must be central and anomalous effects may occur if the specimen is too thick or its edges too irregular.

3 Objectives and eyepieces

The image formed by a microscope is produced by the action of both the objective and the eyepiece. Of the two lenses, the objective is most important because it determines the detail that can be seen in the final image and the optimum useful magnification. It has three important characteristics; namely its magnifying power, its resolving power, i.e. its ability to separate fine detail in the image, and its ability to gather light from the object. The eyepiece enlarges the image formed by the objective to the optimum overall magnification and corrects residual aberrations in the primary image, but it cannot improve the detail in the image formed by the objective.

3.1 Magnification, resolution and perception

3.1.1 *Objective magnification*

The magnification produced by an objective is determined by its focal length and the optical tube length, as described in Chapter 2. For a given tube length the focal length determines the object distance and the primary magnification M, which is given by the simple equation

$$M = t/f \tag{3.1}$$

where t = optical tube length, and f = focal length of the objective.

3.1.2 *Limit of resolution and resolving power*

The limit of resolution is a measure of the ability of the objective to separate in the image individual, adjacent details that are present in the object; it is the distance between two points in the object that are just resolved. The resolving power is the reciprocal of the limit of resolution and is expressed in units of lines resolved per millimetre.

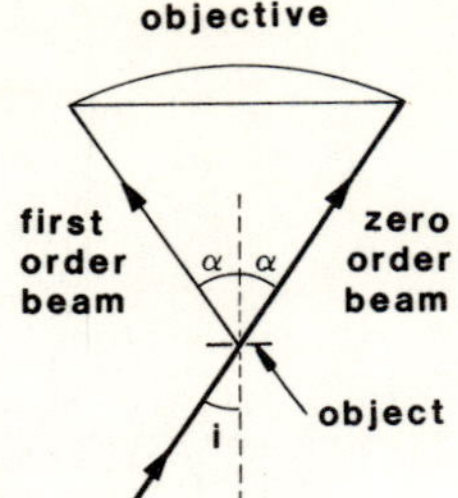

Figure 3.1 Collection of zero- and first-order diffracted beams by the objective.

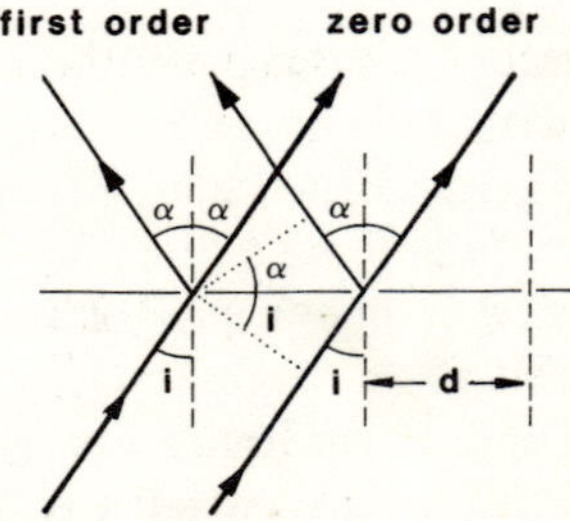

Figure 3.2 The condition for the diffraction of first-order beams by a ruled grating.

In Chapter 2 we saw that for resolution to occur at least both zero- and first-order diffracted beams must be collected by the objective. Suppose a grating with rulings of spacing d is illuminated in vacuum at an angle of incidence i by an oblique beam of coherent, monochromatic light of wavelength λ, and that the zero- and first-order beams are just collected by the objective (Fig. 3.1). Since the path difference between zero- and first-order diffracted rays from successive rulings in the grating is exactly one wavelength (Fig. 3.2),

$$d \sin i + d \sin \alpha = \lambda \tag{3.2}$$

where 2α is the angle through which the first-order beam is diffracted. However, since the two beams are just collected by the objective i is equal to α, thus the limit of resolution, $d_{\min}$, is

$$d_{\min} = \frac{\lambda}{2 \sin \alpha} \tag{3.3}$$

Moreover, if the object space is filled with a medium of refractive index n, the wavelength of the light in the medium, λ_n, is

$$\lambda_n = \lambda / n \tag{3.4}$$

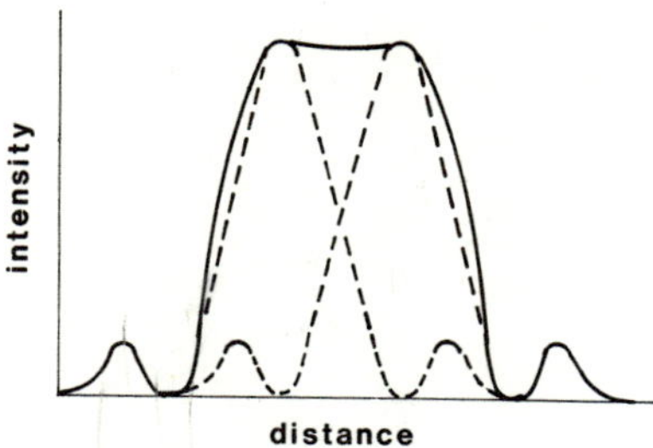

Figure 3.3 Overlapping intensity distributions for the images of two points. The central bright disc in each image coincides with the first dark ring of its neighbour, corresponding to failure of resolution.

Therefore, when an objective is used with the object space filled with a medium of refractive index n,

$$d_{\text{min}} = \lambda_n/2 \sin\alpha = \lambda/2n \sin\alpha = \lambda/2\text{NA} \tag{3.5}$$

$n \sin\alpha$ is called the numerical aperture, NA, of the objective, and is one of its important characteristics.

Lord Rayleigh arrived at the criterion for resolution using a complementary approach. Owing to diffraction, the image of a point of light is a point surrounded by concentric dark and light rings, the intensities of which decrease rapidly. When the images of two points are brought sufficiently close together for their diffraction patterns to overlap, the points cease to be distinguishable. Rayleigh assumed that this occurs when the separation of the points in the image is equal to the radius of the first dark ring, i.e. when the first dark ring of point 1 coincides with the central bright disc of point 2, and vice versa (Fig. 3.3). Quantitatively, this leads to an expression for the limit of resolution

$$d_{\text{min}} = K\lambda/2\,\text{NA} \tag{3.6}$$

where K has a value of 1.22 when the objective is filled with light from a suitable condenser. The value is close to that deduced from diffraction theory, equation (3.5).

From equation (3.5) we see that the resolution can be increased either by increasing the NA or by decreasing the wavelength of the light used to illuminate the specimen. Consequently, microscopes usually are provided with a series of objectives with NA values in the range 0.1 to 1.3; the latter value is close to the practical upper limit for NA. The effect on resolution of increasing the NA from 0.25 to 0.55 is shown in Fig. 3.4 *a* and *b*.

Visible light ranges in wavelength from ~650 nm at the red end of the spectrum to ~400 nm at the blue end. Therefore, the limit of resolution at the blue end is about one and a half times better than at the red end.

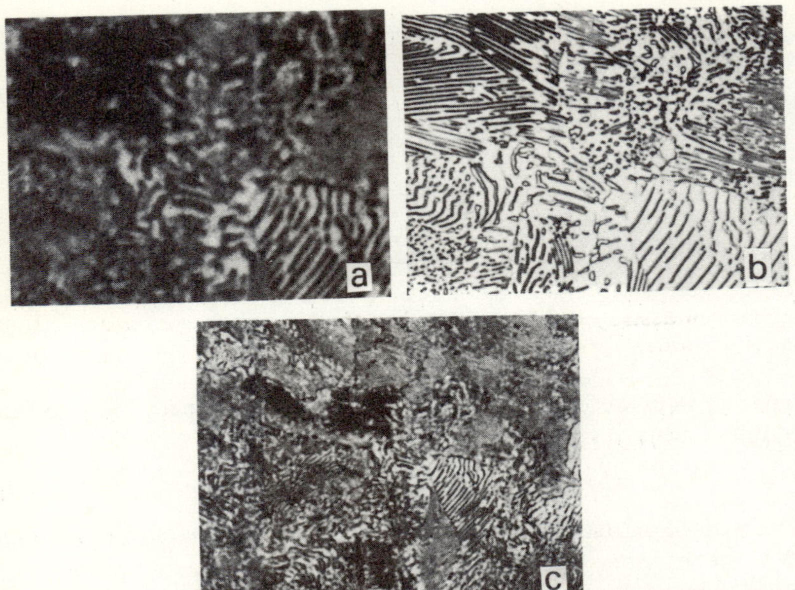

Figure 3.4 Effect of NA on the resolution of pearlite in green light, bright-field illumination. (*a*) NA = 0.25, magnification × 500 (2000 × NA), resolution has failed except at the lower left-hand corner; (*b*) NA = 0.55, magnification × 500 (900 × NA), same field as (*a*), the lamellar structure is largely resolved; (*c*) NA = 0.25, magnification × 200 (800 NA), same field as (*a*). Micrograph (*c*) reveals all the detail visible in (*a*), which illustrates empty magnification.

However, the eye is very insensitive to blue light, so that the improvement in resolution is obtained at the expense of poor visibility. The eye is most sensitive to light in the yellow-green part of the spectrum, i.e. of wavelengths 520–600 nm, and in practice light which is rich in these wavelengths, e.g. daylight or light from a tungsten filament lamp, is often used for visual examination. In this case, the limit of resolution is likely to correspond to that of light of wavelength ~ 560 nm. For photomicrography in black and white, filtered light is used, because essentially monochromatic light gives a sharper image; in practice green light is most often used because objectives are best corrected for aberrations for light from this part of the spectrum.

Further improvement in resolution by up to a factor of two can be achieved by the use of ultraviolet light of shorter wavelengths than visible light. However, focusing ultraviolet light is difficult and optical glasses strongly absorb wavelengths less than ~ 300 nm. Consequently, either objective lenses and eyepieces made from quartz glass, which transmits wavelengths down to ~ 200 nm, or catoptric (mirror) objectives, which also overcome the focusing difficulty, must be used. In practice, the use of ultraviolet microscopes is unusual, because the improvement in resolution which they offer is small compared with that possible in the electron microscope.

3.1.3 *Perception*

Resolution is concerned with the ability of the objective to separate in the image features that are close together. However, features can be perceived when they cannot be resolved. Thus the image of a single pinhole broadens when the NA is small or the wavelength of the light long, because more concentric dark and light rings are formed around the bright disc, but the image is not destroyed. Moreover, when the NA is increased more light is diffracted into the central disc, the contrast is increased and the image is perceived more easily. The theory of perception is complex, but it shows that the limit of perception usually is between a tenth and a hundredth of the limit of resolution. In practice, it is useful to know that a feature is present, even though its detail cannot be resolved.

3.2 Light grasp

The light-gathering power of the objective, or light grasp, is determined by the angle of the cone of rays that the objective can accept, thus it is related to the NA. The angle of the cone depends on the relationship between the diameter of the front lens of the objective and its focal length (see Fig. 3.5*a*); for a given diameter, the shorter the focal length the greater will be the angle and the NA. In practice, NA increases as focal length decreases, but the precise combination of the two depends on the lens design and the need for eyepiece magnification to lie within the range $\times 4$ to $\times 20$.

It is important that the objective should receive as wide a cone of rays from the object as possible, because the angle of the cone of rays affects the

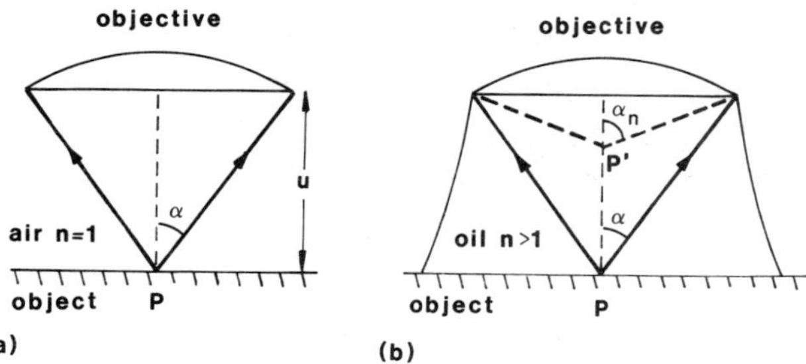

Figure 3.5 Principles of dry and immersion objectives for uncovered specimens. In (*a*) the maximum semiangle of the cone of rays accepted by the objective is α. However, in (*b*) because the refractive index of the oil is > 1, the rays appear to diverge from P', thus increasing the maximum semiangle of the cone of rays to α_n. The effect of the oil is analogous to that of water in a pool, which causes an object on the bottom to appear to be closer to the surface.

brightness of the image, as well as the resolution. Thus for greatest image brightness the objective must be completely filled with light; if the cone of rays has a semi-angle less than α (Fig. 3.5*a*) the image is less bright.

The light grasp is proportional to the area of the front lens of the objective, which is proportional to $\sin^2 \alpha$ and hence to NA^2. Since NA increases with primary magnification, the reduction in intensity accompanying increase in magnification, which spreads the light over an area which is proportional to the square of the magnification, is partially offset by the increase in the light grasp.

This is illustrated clearly by comparing the intensities of illumination in the primary images produced by a $\times$ 10 objective, *A*, with an NA of 0.28 and a $\times$ 40 objective, *B*, with an NA of 0.85. If the NAs of the two objectives were the same, the ratio of the intensities of the images would be

$$\frac{\text{intensity of image } B}{\text{intensity of image } A} = \frac{\times 10^2}{\times 40^2} = \frac{1}{16}$$

However, since the ratio of the light grasps of *B* and *A* is $0.85^2/0.28^2 = 9.2$, the actual ratio of intensities of the images is $9.2/16 = 0.575$. Thus the two images do not differ markedly in intensity, and usually we are not conscious of a change in intensity on changing from one objective to another.

3.3 Dry and immersion objectives

For dry objectives the medium in the object space is air, for which the refractive index is 1.0. Consequently the NA must be less than one; in practice the limiting value is about 0.95, but lenses with NAs exceeding 0.85 are rare. Immersion objectives, which employ either a transparent oil of

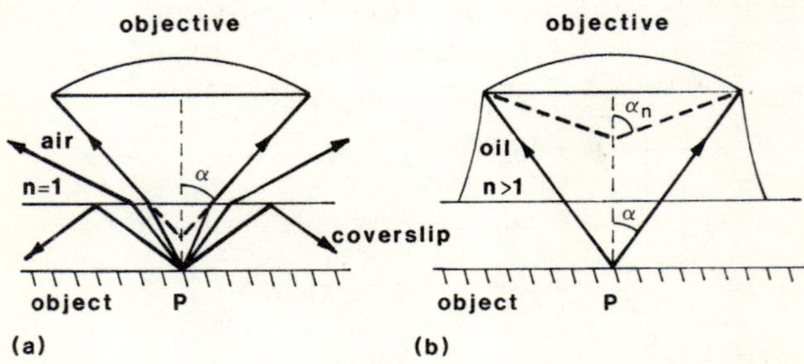

Figure 3.6 Principle of dry and immersion objectives for covered specimens. In (*a*) the rays from point *P* are refracted (1 and 2) or reflected (3) at the surface of the coverslip, thus ray 2 does not enter the objective owing to refraction; the maximum semiangle of the cone of rays accepted by the objective is α. In (*b*), because the refractive index of the oil is about the same as that of the coverslip, rays are only slightly refracted at its surface; moreover because the refractive index of the oil is > 1 the rays appear to diverge from P', increasing the effective maximum semiangle of the cone of rays to α_n.

specified refractive index, or water,* or water and glycerin* in the object space are used because the bigger refractive index of the medium enables values of NA greater than one to be obtained.

The principles of dry and immersion objectives are shown in Fig. 3.5 for uncovered specimens, and in Fig. 3.6 for covered specimens. With uncovered specimens, when the medium filling the object space is air the semi-angle of the cone of rays radiating from an object point P that enters the objective is α, when the free working distance is u (Fig. 3.5*a*). However, assuming the same free working distance, when the medium filling the object space has a refractive index greater than one the rays from the object point P appear to diverge from point P' (Fig. 3.5*b*). This produces an effective increase in the semi-angle of the cone of rays that enter the objective from α to α_n. For specimens with cover slips, when the space between the cover slip and objective is air-filled, the rays from the point P are refracted or reflected at the coverslip-air interface (Fig. 3.6*a*), thus the objective accepts a cone of rays of semi-angle α. However, when the space is filled with immersion oil the rays are deviated to a smaller extent when they cross the interface (Fig. 3.6*b*). Consequently, the objective accepts a cone of rays of greater semi-angle, α_n.

Objectives are designed to be used with a medium of specified refractive index in the object space. Thus dry objectives will not produce satisfactory images when immersion oil occupies the object space and vice versa with oil immersion objectives.

Moreover, objectives for transmitted-light microscopes are designed to work with cover slips on the specimens, while objectives for metallurgical microscopes are designed to work with specimens which do not have cover slips. Consequently, it is important that the two types of objective should not be confused. In practice, low-power dry objectives and oil immersion objectives of the two types can be interchanged without severe loss of image quality, but interchange of medium- and high-power dry objectives produces badly degraded images.

Traditionally, cedar wood oil was used as the immersion fluid, but had the disadvantage that it dried, i.e. polymerized, if left on the objective, and was difficult to remove. Modern synthetic immersion oils are non-drying; nonetheless after use they should be removed from the lenses with lens cleaning tissue.

Using immersion oil of refractive index ~ 1.52, the maximum semi-angle of the cone of rays that can be accepted by the objective is $\sim 67°$, corresponding to a maximum NA of ~ 1.4, although the majority of oil immersion objectives have NAs ~ 1.3. The maximum value is limited by the

*These immersion media are not usually used in work on materials.

working distance of the objective and the angle for total reflection.

When using oil immersion objectives on a transmitted-light microscope it is necessary to oil both the objective and the substage condenser to the specimen. Moreover, the condenser must have a sufficiently large NA to enable the maximum resolution of the objective to be achieved. This problem does not arise with immersion objectives on metallurgical microscopes because the objectives act as their own condensers.

Many ceramic and petrological specimens are mounted in a transparent medium. To avoid total internal reflection in the specimens when using oil immersion objectives, and the loss of the advantage of such objectives, the mounting medium employed must have a high refractive index. Canada balsam, a medium commonly used, has a refractive index of ~ 1.54, and satisfies this criterion.

3.4 Depth of field and field of view

When a lens is sharply focused on a given plane, points which lie a little further from or closer to the objective also appear in acceptable focus. The range of distances over which points appear in acceptable focus is called the *depth of field*. Sometimes it is referred to incorrectly as the depth of focus*. Thus the object plane is really a thin layer in which all points appear to be in focus. Qualitatively, this occurs because points in the image are Airy discs, not perfect discs of light. Consider two parallel planes P_1 and P_2 which are close together and perpendicular to the optic axis of the objective (Fig. 3.7), and suppose that the objective is sharply focused on plane P_1, producing an image at I. Rayleigh suggested that for plane P_2 also to appear just in focus the paths in the image space of the marginal rays from the two planes should differ by less than a quarter wavelength. If the difference is greater the bundle of rays from a point P_2 produces an image point that is greater in diameter than the circle of confusion arising from diffraction. Using this

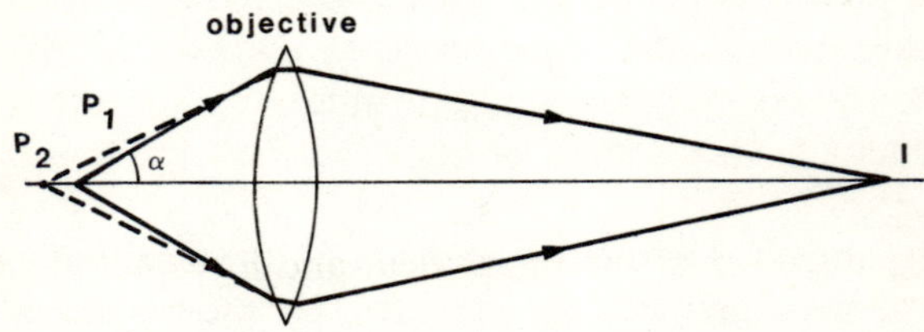

Figure 3.7 The principle of depth of field of an objective.

* The depth of focus is correctly defined as the distance through which the image plane may be moved before the image becomes unsharp.

Table 3.1 Effect of numerical aperture on depth of field

Numerical aperture		*0.10*	*0.25*	*0.50*	*0.75*	*1.00*	*1.25*
Depth of field	Air, $n = 1.0$	56	8.6	2.0	0.81	(0.27)	—
(μm).	Oil, $n = 1.5$	82	13	3.2	1.4	0.73	0.42

Wavelength of light 550 nm.

criterion the depth of field, h, can be shown to be

$$h = \pm(\lambda/4)/n\sin^2(\alpha/2) = \pm n\lambda/\mathrm{NA}^2 \tag{3.7}$$

where $\lambda/4$ = the maximum allowable path difference which determines the depth of field, and n = the refractive index of the object space. In practice, the equation gives a conservative estimate of the depth of field. The calculated effect of NA on the depth of field for dry and immersion objectives is shown in Table 3.1.

The depth of field is limited, especially with objectives of high NA. This is important for three practical reasons. Firstly, focusing becomes more difficult the smaller the depth of field. Accordingly, with a series of parfocal objectives it is good practice to focus first with the objective of lowest NA, because there is less chance of missing the plane of focus. Secondly, it is difficult to examine objects with irregular surfaces, e.g. fracture surfaces, at higher magnifications because it is impossible to obtain all the surface in focus simultaneously. Thirdly, mounting the surface to be examined perpendicular to the optic axis of the microscope becomes more critical the smaller the depth of field.

The *field of view* depends on the characteristics of both objective and eyepiece. In practice, the image formed by the objective is formed about 18 mm below the rim of the microscope tube. Consequently with a microscope tube of standard internal diameter, 23.2 mm, the size of the primary image is restricted. The design, correction and focal length of the eyepiece determine the diameter of the primary image that can be observed; its maximum value is 18–19 mm. This diameter, measured in mm, is described as the field-of-view index of the eyepiece, S.

From the simple theory of geometrical optics

$$m = v/u = d_i/d_0 \tag{3.8}$$

where d_i = the diameter of the image field and d_0 = the diameter of the object field. Hence, since $d_i = S$,

$$d_0 = S/m \tag{3.9}$$

For example, if the field-of-view index is 18, the diameters of the fields of view of × 40 and × 10 objectives are 0.45 and 1.8 mm respectively. In

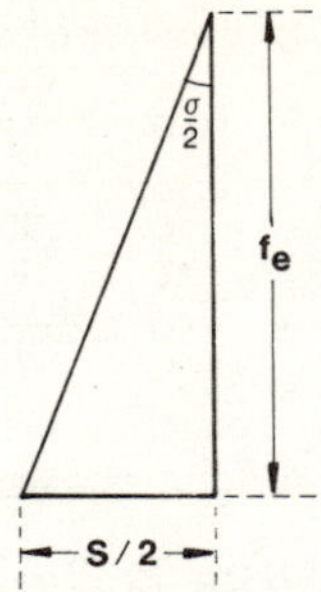

Figure 3.8 The relationship between the diameter of the primary image and the angle of view.

general, the lower the magnification of the objective the greater becomes the field of view.

Moreover, the field of view index and the focal length of the eyepiece, f_e, determine the angle of view, σ, i.e. the angle of the cone of rays entering the eye. This angle (Fig. 3.8) is given by the equation

$$\tan(\sigma/2) = S/2f_e \tag{3.10}$$

Thus for a $\times$ 6 eyepiece with $S = 18$, when the image is formed at the nearest distance of distinct vision, 250 mm, $f_e = 250/6 \simeq 42$ mm and $\sigma \simeq 25°$; typically for Huyghenian eyepieces σ is in the range 25° to 35°.

The eye can accommodate an angle of view up to $\sim 50°$. Accordingly, many microscope makers offer wide-field eyepieces, with large-field tubes of about 30 mm diameter, for use with flat-field objectives and microscope tubes of greater diameter than standard.

3.5 Optimum useful magnification

To make proper use of an objective we need to know how much the image must be magnified so that we can perceive all the detail in it. This depends on the ability of the eye to resolve detail. The unaided eye is able to just distinguish between two points if they subtend an angle of one minute of arc at the eye, which corresponds to a separation of $\sim 100\,\mu$m at the nearest distance of distinct vision. However, for working without eye strain the separation needs to be two to four times larger.

The minimum magnification needed for the eye to just distinguish between adjacent points is the ratio of the minimum separation of two points that can be just distinguished by the eye to the resolution of the objective. Thus, we compare a low-power dry objective and a high-power oil immersion objective in Table 3.2, assuming the use of yellow light.

Comparison of the data for the two objectives shows that the ratio of

Table 3.2 Estimation of the overall magnification required to enable resolved detail to be seen

	Low-power dry objective	*High-power oil-immersion objective*
Wavelength of light	580 nm	
Numerical aperture	0.15	1.30
Focal length	16 mm	1.8 mm
Objective magnification	× 10	× 95
Resolution ($\lambda/2$ NA)	1935 nm	223 nm
Minimum magnification	$10^{-4}/1.935 \times 10^{-6}$ = 52.5	$10^{-4}/2.23 \times 10^{-6}$ = 460
Minimum eyepiece magnification	$52.5/10 \simeq 5$	$460/95 \simeq 5$
Optimum magnification (× 2 to × 4 minimum magnification)	100 to 200	1000 to 2000
Optimum eyepiece magnification	10 to 20	10 to 20
Optimum magnification/NA	660 to 1300	770 to 1500

optimum magnification to objective NA is about 1000, and a useful rule of thumb is that the overall magnification should be 500 to 1000 times the objective NA. Moreover, both require the use of eyepieces of the same magnifying power to achieve the optimum magnification; this is generally true for all objectives.

While it is necessary for the overall magnification to exceed a minimum value if we are to see all the resolved detail, excessive magnification should be avoided. It increases the size of the image without increasing the amount of detail and may degrade its quality; it is called empty magnification (Fig. 3.4*a* and *c*).

3.6 Defects in lenses

The quality of the final image is no better than that of the primary image produced by the objective. In practice, there are several defects that can mar the quality of the image produced by a lens, especially an objective, the most important of which are spherical aberration, chromatic aberration, astigmatism, coma, curvature of field, and distortion: they are considered below.

3.6.1 *Spherical aberration*

When monochromatic light rays from a point source pass through the outer parts of a lens they are brought to a focus at a shorter distance from the lens than are those that pass through the central parts (Fig. 3.9), owing to the more oblique rays being refracted through a greater angle than those near the optic axis. The aberration is caused by the spherical shape of the

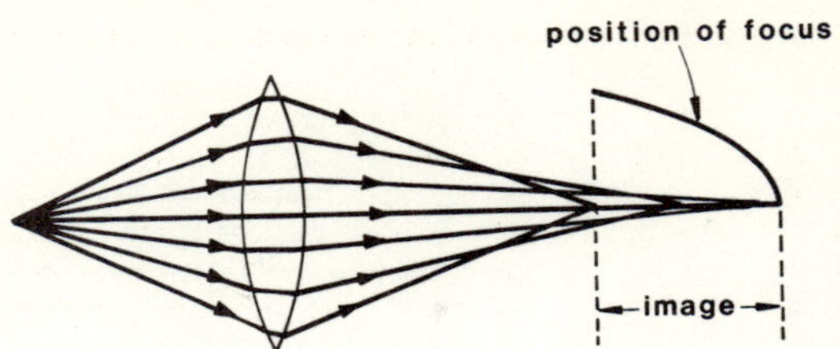

Figure 3.9 Spherical aberration. Rays passing through different lens zones do not focus at a point. The curve indicates the amount of spherical aberration.

lens surfaces, hence spherical aberration, and it is more severe the greater the aperture of the lens. It occurs for most positions of an axial object point, but for certain positions it becomes zero. Such aberration-free object and image points are called aplanatic points. For a spherical surface one pair of such points lies at distances *nr* and *r*/*n* from the centre of curvature, where *r* is the radius of curvature. This property is used in the design of high-power objectives, the aplanatic image point of the first spherical surface being made the aplantic object point of the next lens. This correction for spherical aberration is effective only at a fixed tube length, because altering the tube length alters the positions of the object points. Moreover, the aberration can be largely, but not completely, eliminated by use of combinations of converging and diverging lenses of different refractive index. These corrections are also effective for coma, which is a form of asymmetrical spherical aberration.

Objectives are designed for use with either specimens with glass cover slips 0.17 mm thick, or specimens without cover slips. Omission or introduction of a cover slip as the case may be upsets the correction for spherical aberration. Thus for transmitted-light microscopes the cover slip is a component of the optical system, but not for the metallurgical microscope. The change in the optical path caused by a cover slip of incorrect thickness is shown in Fig. 3.10. Optically the cover slip is equivalent to a plane parallel plate which deflects the rays more strongly the

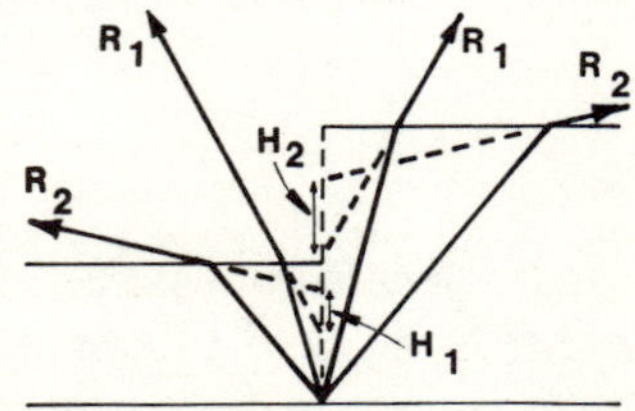

Figure 3.10 Effect of coverslip thickness on the optical paths of rays from a point on a specimen.

more they diverge from the perpendicular. Thus with a cover slip of standard thickness, 0.17 mm (Fig. 3.10, LHS), the rays R_1 and R_2 appear to originate at two points distance H_1 apart, whereas with a thicker cover slip (Fig. 3.10, RHS), they appear to originate at points a greater distance H_2 apart, which impairs the image quality. With dry objectives the required accuracy of the thickness of the cover slips increases with increase in the aperture of the objective. However, with oil immersion objectives any variation in thickness is largely compensated by change in the thickness of the layer of oil, because the refractive indices of glass and oil are almost identical. Consequently the image quality is much less sensitive to the thickness of the coverslip. Sometimes objectives are fitted with correction mounts that allow compensation to be made for variations in the thickness of the cover slip.

3.6.2 *Astigmatism and curvature of field*

Astigmatism is a defect in which the images of points off the optic axis are drawn out into blurred lines or discs. It increases with distance from the optic axis and causes poor definition of images formed away from the axis. The extent of the aberration can be reduced by stopping down the lens, but this impairs resolution. The cause of astigmatism is illustrated in Fig. 3.11; the tangential plane is defined by the vertical diameter, TT', of the lens, while the sagittal plane is defined by the horizontal diameter, SS'. The tangential rays OT and OT' produce the tangential image, I_t, and the sagittal rays OS and OS' produce the sagittal image, I_s, that lie on surfaces of different curvature. The horizontal line image, I_t, lies in the sagittal plane

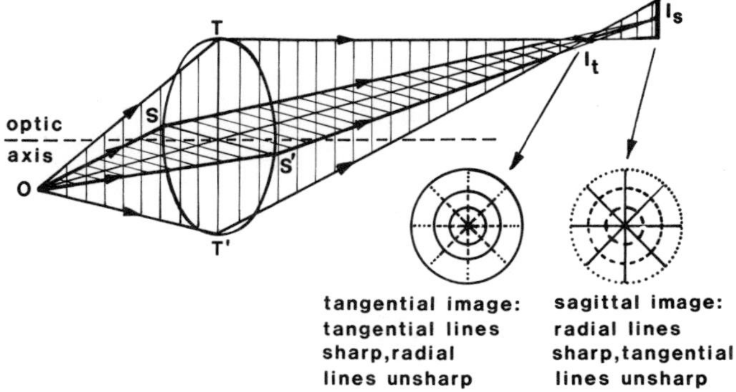

Figure 3.11 Astigmatism. Rays lying in the tangential plane are brought to a different focus from those lying in the sagittal plane. In image I_t tangential lines are focused, while in image I_s radial lines are focused. Best definition for peripheral detail is obtained in a plane midway between I_t and I_s.

and the vertical image I_s, in the tangential plane. The circle of least confusion lies between the two images and is the best image possible. Correction for astigmatism causes the two images to coincide, but they continue to lie on a curved surface.

Curvature of the image field arises from the change in focal length of the lens as the position of the point on the lens moves away from the optic axis; it depends on the lens geometry and refractive index. The radius of curvature of the field is approximately equal to the focal length of the objective. Consequently it is more apparent in images formed by high-power objectives. It cannot be eliminated in objectives of conventional design, but is largely eliminated in modern flat-field objectives. With conventional objectives it is partially overcome by the use of suitable negative (Huyghenian) compensating eyepieces, because corrections to the objectives lead to sacrifice of the extreme sharpness of the image. Compensation for field curvature is particularly important in photomicrography, where a large field which is sharply in focus is required.

3.6.3 *Coma and distortion*

Coma causes the image of a non-axial point to be reproduced as an elongated comet shape, lying in a direction perpendicular to the optic axis. It is a form of asymmetrical spherical aberration affecting non-axial object points. The comet is made up of a series of overlapping circular images (Fig. 3.12) that increase in size in going from I_1 and I_3. It results from differences in magnification arising from rays meeting the lens at widely differing angles. The overlapping images decrease in intensity from I_1 and I_3. As with astigmatism, it can be reduced by stopping down the objective. Correction is achieved by figuring the lens surfaces so that the ratio

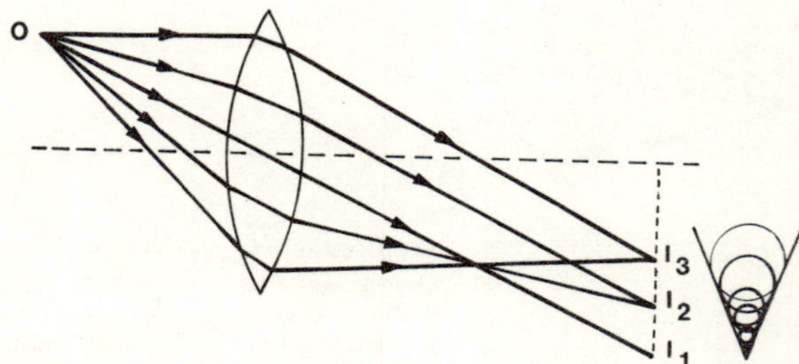

Figure 3.12 Coma. Oblique incident rays which pass through the outer parts of the lens do not converge to the same point as those passing through the central parts. The image of a non-axial points consists of a series of overlapping circular images which increase in size but decrease in intensity from I_1 to I_3.

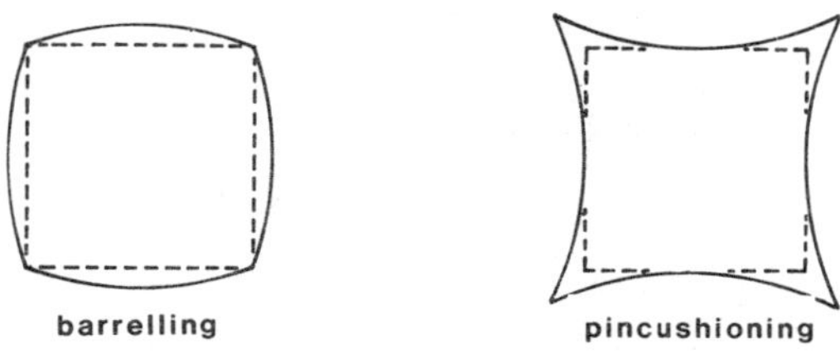

Figure 3.13 Distortion. (*a*) Barrelling and (*b*) pincushioning, owing to variation in magnification with distance from the optic axis.

sine (angle incident ray)/sine (angle emergent refracted ray) is constant. Modern objective lenses are usually free from coma.

Distortion arises from variation in magnification with distance of the object point from the optic axis. It causes either barrelling or pincushioning (Fig. 3.13), depending on whether the magnification decreases or increases with distance from the optic axis. It occurs in both objectives and eyepieces, but is more common in the latter, and is difficult to eliminate completely.

3.6.4 *Longitudinal chromatic aberration*

The aberrations already discussed occur when monochromatic light is used, but additional aberrations arise when the light is not monochromatic. Thus, when white light is focused by a lens, light of different wavelengths is brought to focus at different distances from the centre of the lens (Fig. 3.14), violet light being focused closer to the lens than red light. Consequently a simple lens produces several images of the object in different colours, each focused in a slightly different plane, causing the image to have coloured fringes around its outline.

This occurs because the refractive index of a transparent isotropic material is greater for light of shorter wavelength than for light of longer wavelength. The proportion in which the refractive indices differ is called the *dispersion* of the material. Strictly, it is measured by the slope of the

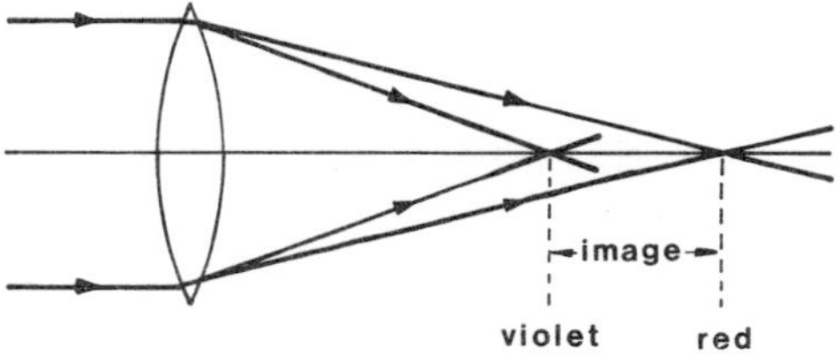

Figure 3.14 Longitudinal chromatic aberration. Light of shorter wavelengths is brought to focus closer to the lens than light of longer wavelengths, causing the image to be surrounded by colour fringes.

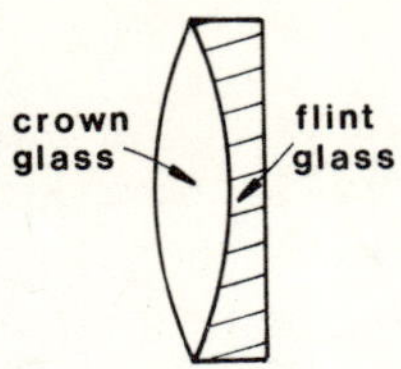

Figure 3.15 An achromatic doublet.

curve of refractive index versus wavelength, $(dn/d\lambda)$, and is greater for shorter wavelengths than for longer wavelengths. However, in optical calculations a quantity called the reciprocal relative dispersion, ν, is used; this is defined as

$$\nu = (n_D - 1)/(n_F - n_C) \tag{3.11}$$

where n_D is the refractive index of the sodium D line (589.3 nm), n_F the refractive index of the hydrogen F line (486.1 nm), and n_C the refractive index of the hydrogen C line (656.3 nm).

Different kinds of glass have different reciprocal relative dispersions, e.g. crown glass ~ 60 and flint glass ~ 38. Consequently a converging lens of crown glass can be combined with a weaker diverging lens of flint glass (Fig. 3.15), so that the chromatic aberrations cancel for certain wavelengths, while leaving some converging power: the combination is called an achromatic doublet.

3.6.5 *Lateral or transverse chromatic aberration*

This occurs because the magnification produced varies with the wavelength of the light. The violet image is larger than the red image, even when the longitudinal chromatic aberration is eliminated and consequently the image is surrounded by weak colour fringes. It is most satisfactorily eliminated by the use of cemented lenses in the eyepieces.

3.7 Types of objective

Lenses that are completely corrected for spherical aberration over the entire field and for coma over the central two-thirds of the diameter, the remainder being almost free from coma, are described as being aplanatic. In addition microscope objectives are also corrected to a greater or lesser degree for chromatic aberration. Thus there are four types of objectives which vary in their degree of correction; these are achromats, fluorites or semi-apochromats, apochromats, and flat-field objectives.

3.7.1 *Achromats*

These lenses are corrected for longitudinal chromatic aberration usually in red and blue light and for spherical aberration in green light, as well as for the other aberrations described in the previous section, by use of lenses made from materials of various refractive indices. Consequently, all rays within a range of wavelengths from about 500 nm to about 630 nm are brought to focus within the useful depth of focus (Fig. 3.16). They are well corrected for the range of wavelengths to which the eye is most sensitive and are particularly good for visual examination of specimens, but high-power achromats must be used with compensating eyepieces. For photomicrography using black and white film they should be used with a green filter to permit the lenses to work in the wavelength range for which correction is best and to exclude blue/violet light, for which the lenses are not corrected and to which photographic film is rather sensitive. They are widely used for metallographic work.

3.7.2 *Fluorites or semi-apochromats*

Fluorites are corrected for chromatic aberration to a greater degree than achromats, but to a lesser degree than apochromats. The slight residual aberration which is present in achromats is reduced by the use of fluorspar-like materials, which have much lower refractive indices and dispersions than optical glasses, for some components of the lenses. This extends the

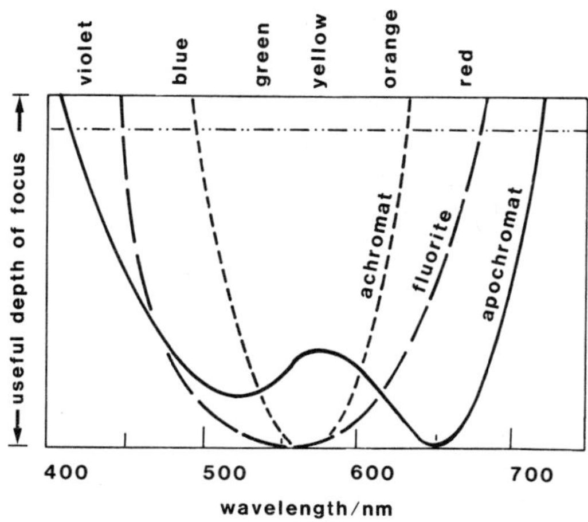

Figure 3.16 Typical colour curves, showing the range of wavelengths which contribute to image formation and the relative positions at which the rays of various wavelengths focus for high power achromats, fluorites and apochromats (after Cooke, Troughton and Simms Ltd.).

range of wavelengths in which all rays are in focus to about 450 to 650 nm (Fig. 3.16), and increases the colour contrast during visual observation, but also introduce severe curvature of field. Consequently, compensating eyepieces must always be used with this type of objective to compensate for this curvature; nonetheless the diameter of field of high-power objectives is severely restricted. Again, they are widely used for metallographic work.

3.7.3 *Apochromats*

In these lenses correction for chromatic aberration in the red, green and blue parts of the spectrum and spherical aberration in two colours is made using combinations of positive fluorite-type lenses with negative lenses of special optical glasses. This almost completely eliminates residual aberrations and produces an image which is sharp and of high contrast. The range of wavelengths that are brought to the same focus is about 420 to 720 nm (Fig. 3.16). It is essential to use compensating eyepieces with these lenses, because the objectives are undercorrected for colour in order to achieve better correction of the spherical aberrations.

3.7.4 *Flat-field objectives*

Objectives of this type fall into two groups, planachromats and planoapochromats. They both produce wide flat-field images, accompanied by a degree of correction for spherical and chromatic aberration that is comparable to that in achromats and apochromats respectively. The improvement in performance is achieved at the expense of many additional components in the lenses. Thus comparing objectives with a magnifying power of × 100 and a numerical aperture of 1.3, typically an achromat would have six, an apochromat ten, and a planapochromat fourteen components. This type of objective is particularly suitable for photomicrography because of the flatness of the image it produces.

3.7.5 *Control of glare in optical components*

Even when light falls normally on an air-glass or glass-air interface some of the light is reflected. In a microscope there are many interfaces at which partial reflection occurs, causing degradation of the image quality by the reflected, non-image-forming light, called glare or flare, and appreciable loss in the intensity of the transmitted beam. For *normal* incidence at an interface between two transparent media the fraction of the incident light reflected, R, is given by the equation

$$R = [(n_1 - n_2)/(n_1 + n_2)]^2 \qquad (3.12)$$

where n_1 and n_2 are the refractive indices of the two media on either side of the interface. Thus a single interface between air and crown glass, $n = 1.517$, reflects 4.2% of the incident light; in general, depending on their refractive indices, optical glasses reflect between 3 and 8% of the incident light at each interface. However, the loss at glass–glass interfaces is much smaller, of the order 0.1%.

The fraction of the incident beam transmitted at an interface is $(1 - R)$ and, if there are x interfaces the fraction of light transmitted is $(1 - R)^x$. Thus assuming a transmission of 0.95 for each interface, seven interfaces reduce the intensity of the transmitted light by 30%. In practice, microscopes may have up to about twenty air-glass interfaces, depending upon the type of objective, consequently both the increase in glare and the loss in intensity of the transmitted beam may be substantial if steps are not taken to reduce the amount of reflection. Moreover the problem of glare is greatly exacerbated in metallurgical microscopes because the light passes through the objective twice.

The amount of light reflected at a glass-air interface can be markedly reduced by coating the surface of the glass with a very thin film of a material with a suitable refractive index; this is sometimes called blooming. This is achieved by causing the fractions of light reflected at the air-coating and coating-glass interfaces to be equal and then arranging for the reflected rays from the two interfaces to interfere destructively.

We can deduce the refractive index of the coating needed to satisfy the first of these conditions as follows. Let the refractive indices of air, coating, and glass be 1.00, n_2, and n_3. For equal fractions of light to be reflected at each interface

$$[(1 - n_2)/(1 + n_2)]^2 = [(n_2 - n_3)/(n_2 + n_3)]^2 \tag{3.13}$$

thus

$$n_2 = \sqrt{n_3} \tag{3.14}$$

i.e. the refractive index of the coating must be equal to the square root of the refractive index of the glass. Moreover, the result is unaltered if multiple reflections occur at the interfaces.

The second condition is satisfied if the two reflected rays are out of phase by exactly half a wavelength; accordingly the path length ABC (Fig. 3.17), must be half a wavelength. For monochromatic light at normal incidence to the interfaces the thickness of the coating, t, must be equal to a quarter of the *wavelength in the coating*, i.e.

$$t = \lambda/4n_2 \tag{3.15}$$

where λ is the wavelength of the light in vacuum. When the condition for

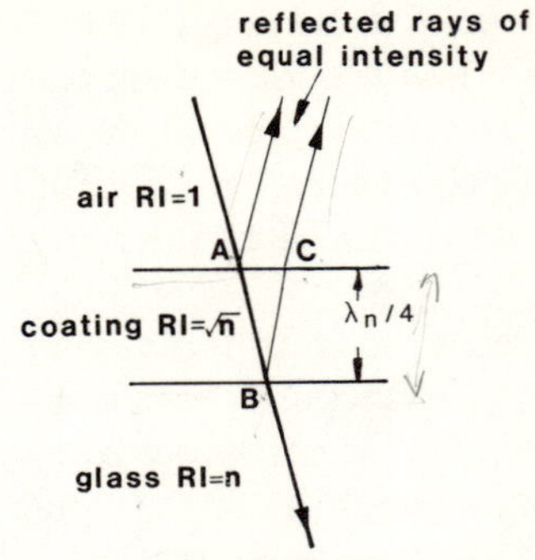

Figure 3.17 The principle of blooming.

destructive interference is fulfilled the energy of the reflected waves is not destroyed but augments the transmitted rays.

For optical glasses the refractive indices of the coatings need to lie within the range 1.22 to 1.29, depending on the refractive indices of the glasses. No coating material has a refractive index in this range and in practice, the most suitable material is magnesium fluoride, with a refractive index of 1.38. Thus the condition for reflected rays of equal intensity is not fully satisfied. Clearly, it is not possible to satisfy the condition for destructive interference for the whole range of wavelengths in white light, and the coating thickness is arranged to satisfy the condition for a wavelength in the middle of the spectrum, i.e. for green light. Consequently a single coating often appears purple or magenta in colour.

In practice, a single coating reduces the amount of light reflected at an air-glass interface to about 1%. Thus in a high-magnification system with twenty air-glass interfaces, coating increases the proportion of image forming light from 36% to 82% of the incident light and the ratio of image-forming light to light producing glare from 0.56/1 to 4.55/1.

Recently, multicoatings have been developed commercially in which up to eleven or more layers are deposited at each interface. They reduce the reflection at an air-glass interface to about 0.1%, which yields more contrasty images and increases the transmission of the lenses.

The coatings are evaporated onto the surfaces under vacuum, sometimes under a high electrical potential. Originally, materials having high vapour pressure at low temperatures were evaporated by electrical heating, but electron beam heating techniques are now used, enabling materials of much lower volatility to be employed as coatings.

3.7.6 *Catoptric or reflecting objectives*

Microscope objectives may be made from curved mirrors instead of from lenses, and have two theoretical advantages. Firstly, they may be designed

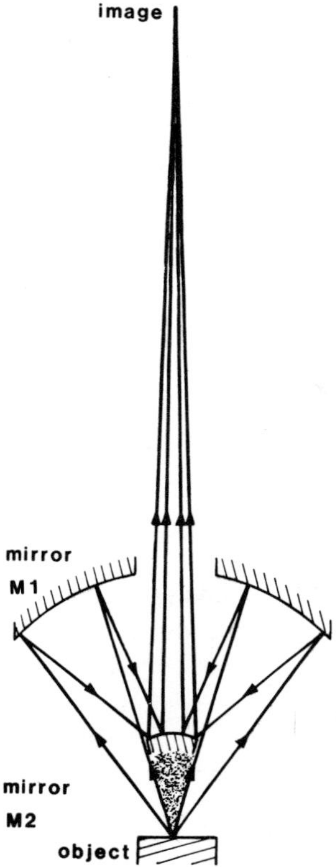

Figure 3.18 Principle of the Cassegrain, or Newton, catoptric objective.

to have much longer working distances than conventional dioptric objectives. Secondly, they bring rays of all wavelengths to the same focus, thus achieving full achromatic correction and enabling the use of ultraviolet and infrared light, both of which are focused with visible light.

In the materials field they are sometimes used for hot-stage or cold-stage microscopy, where advantage is taken of their longer working distance to help keep the objective close to room temperature.

There are two types of reflecting objective, the Cassegrain or Newton type, and the Schwarzschild type. In the Cassegrain type rays of light from the object first fall on a convex mirror of small radius of curvature (Fig. 3.18), and are reflected onto a concave mirror of larger radius of curvature, which focuses them in the image plane. However, in the Schwarzschild type the mirrors are arranged in the opposite sequence (Fig. 3.19). In both cases, part of the cone of rays from the object is stopped off by the smaller mirror as shown by the shaded areas in the figures. The effect is less severe in the Schwarzschild-type objectives and therefore they are usually preferred.

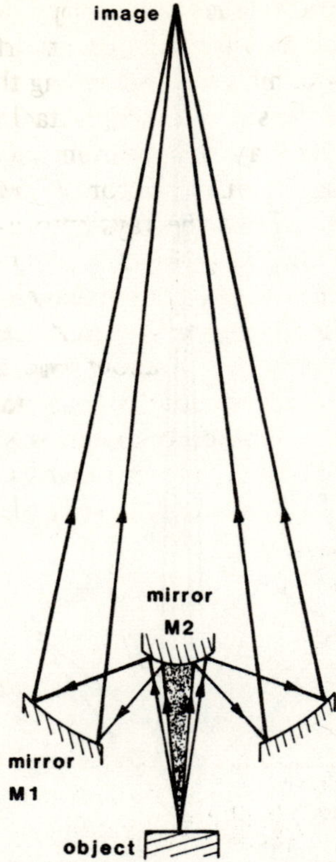

Figure 3.19 Principle of the Schwartzchild catoptric objective.

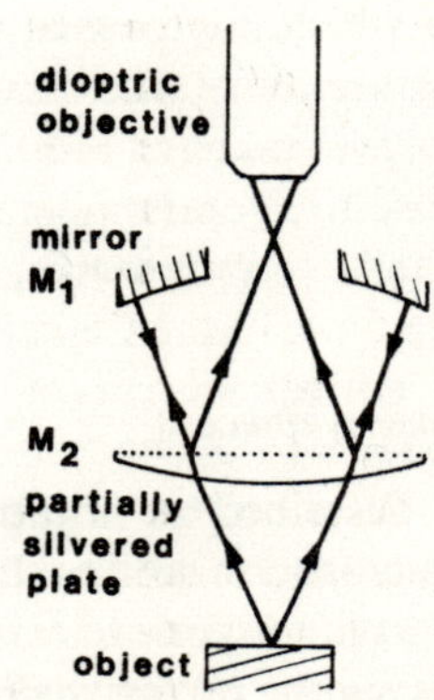

Figure 3.20 Principle of the Dyson long-working-distance attachment.

They are used in the same way as dioptric objectives, but are bulkier and heavier. Thus it is preferable to use them on inverted microscopes.

An alternative method of markedly increasing the working distance of dioptric objectives is the Dyson reflecting attachment, the principle of which is shown in Fig. 3.20. Rays of light from the object pass through a semi-silvered plate, M_2, onto a convex mirror, M_1. M_1 reflects the rays back onto M_2 which partially reflects the rays into a dioptric objective. In practice, the lower side of plate M_2 is made slightly convex to compensate for spherical aberration introduced by the thickness of the plate and by the window in a hot stage. The attachment is reported to increase the working distance of a 4 mm objective from about one millimetre to thirteen millimetres. Because the light is partially reflected at M_2, illuminating systems for bright-field illumination of opaque specimens are difficult to design, and the attachment is used only for examination of such specimens by dark-field illumination. However, it is suitable for examining transparent specimens by transmitted light.

3.8 Eyepieces or oculars

Eyepieces have three main functions; to examine and magnify the primary image, to correct residual lateral chromatic aberration and, especially for photomicrography, to flatten the field of view and correct astigmatism.

For visual examination there should be sufficient clearance between the eye and the eye lens when the eye is placed at the exit pupil of the eyepiece, which is the best position for viewing the entire image. If the eye is at a greater distance from the eye lens the field of view is restricted and the eye must be moved if the whole field is to be examined.

A single-lens eyepiece would produce a cone of rays with an angle greater than that which the eye can accept, so that the eye would be unable to view the entire field simultaneously. Consequently, all eyepieces consist of two lenses, the field lens and the eye lens. There are several types of eyepiece, depending on the degree of correction required and the uses to which they are to be put. The types are Huyghenian, compensating, Ramsden or Kellner, wide-field and projection eyepieces; they are described below.

3.8.1 *Huyghenian or negative eyepieces*

A Huyghenian eyepiece is described as negative because it produces a reduced image when used as a hand magnifier. It is the most simple kind of eyepiece and consists of two plano-convex lenses of crown glass with the convex sides of the lenses facing the objective (Fig. 3.21). The focal lengths of the field and eye lenses are in the ratio 1.5 : 1 to 3 : 1 depending on the

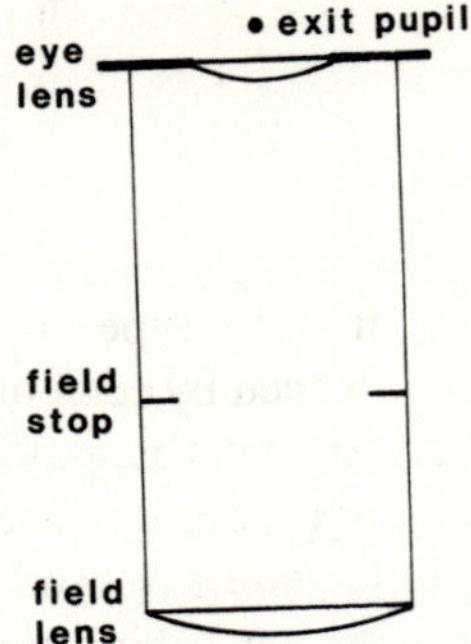

Figure 3.21 The Huyghenian eyepiece.

magnifying power of the eyepiece, and the distance between the lenses is about half the sum of their focal lengths, to eliminate chromatic differences of magnification. The field lens is arranged to lie a little in front of the primary image plane in order to reduce the obliquity of the marginal rays (Fig. 3.22). Consequently, the primary image is formed behind the field lens and is slightly diminished in size. A circular diaphragm, the field stop (Fig. 3.21), is placed in the plane of this image to give a clean edge to the field and, if required, it can be used to support a graticule (reticule).

Some spherical and chromatic aberrations occur and the field of view is small, its angle rarely exceeding 35°, and is slightly curved. Nonetheless, when used with low- or medium-power achromats Huyghenian eyepieces produce images free from colour fringes. However, they are unsuitable for use with fluorites, apochromats and flat-field objectives, which require more highly corrected eyepieces.

A serious limitation is the small eye clearance, i.e. the exit pupil is very close to the eye lens, which makes eyepieces with magnifying powers greater than × 20 impractical. Moreover, because the eyepieces become very long and their field of view restricted when their magnifying power is less than × 4, the preferred range of magnifying powers is × 5 to × 12. The small eye

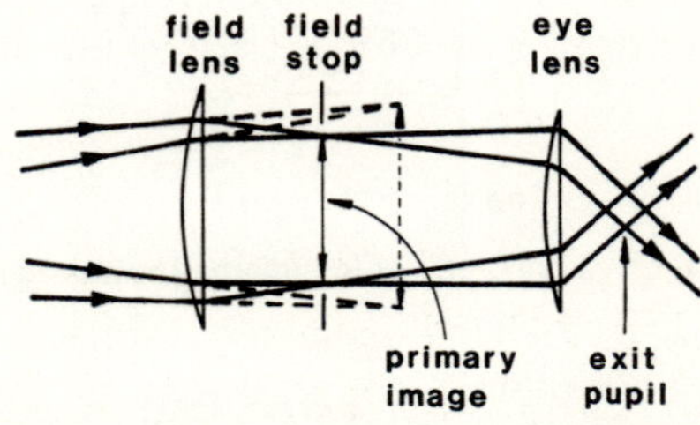

Figure 3.22 Diagram showing the effect of the field lens in reducing the obliquity of the marginal rays in the Huyghenian eyepiece.

clearance also creats difficulties if spectacles must be worn when using the microscope*.

3.8.2 *Compensating eyepieces*

These are usually of the Huyghenian type, in which the eye lenses and sometimes the field lenses are replaced by cemented doublets to achieve the desired corrections. Lateral chromatic aberration, which occurs especially in the marginal regions of the image, is eliminated, giving images free from colour fringes, and some field flattening is produced. Thus, compensating eyepieces should always be used with high-power achromats, fluorites, apochromats, and flat-field objectives. Occasionally, high-power compensating eyepieces are of the positive type to give greater eye clearance.

3.8.3 *Ramsden and Kellner or positive eyepieces*

These eyepieces are described as positive because they form a magnified image when used as hand magnifiers. They consist of two plano-convex lenses, with the convex sides of the lenses facing each other (Fig. 3.23). The primary image is arranged to lie beneath the field lens in the plane of the field stop. Because the primary image is not affected by the field lens, they are especially useful for making measurements and comparisons with graticules (reticules). Moreover, because the field stop is beneath the field lens, fitting and interchange of graticules is easy. Unlike the Huyghenian eyepieces, they have large eye clearance. However, simple Ramsden eyepieces suffer from considerable oblique chromatic aberration and, consequently, the eyepiece is achromatized by making the eye lens a

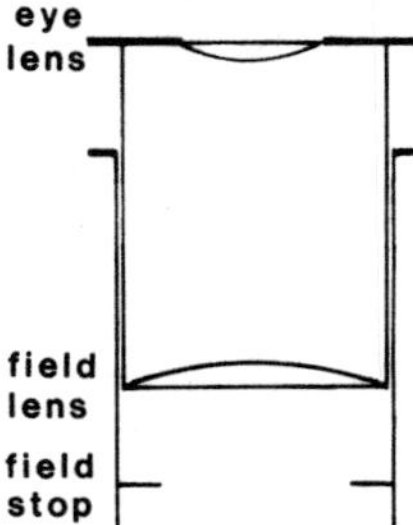

Figure 3.23 The Ramsden eyepiece. The eyepiece unit may be moved relative to the field stop to permit focusing on the plane of the stop.

* Special long eye clearance eyepieces, which have exit pupil distances of about 20 mm, are manufactured for spectacle users who are unable to dispense with their spectacles when using the microscope.

doublet; it is then described as a Kellner eyepiece. When the eyepiece is intended for the use with a graticule the eyepiece is made adjustable so that the graticule can be sharply focused.

3.8.4 *Projection or amplifying eyepieces*

For photomicrography a real image must be produced at a finite distance. Using one of the types of eyepiece discussed previously, this may be achieved by refocusing the microscope so that the primary image falls in front of the focal plane of the eye lens. However, this reduces the optical tube length of the microscope and the image formed may show marked curvature of field, which is usually unacceptable. These difficulties are overcome by the use of specially designed projection eyepieces, that usually are of the negative type, and which may be equipped with focusing eye lenses. For recording images on 35 mm film some manufacturers now offer microprojection eyepieces, which have magnifying powers in the range × 1 to × 5; with this size of film part of the magnification is produced in the photographic enlarging process.

3.8.5 *Widefield eyepieces*

Accompanying the development of flat-field objectives has been the development of wide-field eyepieces; these are of the compensating type, but cover a field of view up to 50% greater than conventional compensating eyepieces. In order to achieve this and to maintain the corrections over a much wider field of view, the eye lens becomes more complex.

4 Illumination of the object

Few objects are self-luminous, hence it is usually necessary to illuminate the object. The illumination is of great importance because it affects the quality of the image formed by the microscope; the microscope cannot make good deficiencies in the illumination.

An illuminating system must fulfil three criteria. It must allow maximum resolution and contrast to be obtained in the image, it should be simple and easy to adjust and, especially for photomicrography, the field of view must be uniformly illuminated.

An objective gives maximum resolution and contrast if the cone of light from the object just fills the objective and if light enters the objective only from the field observed. Illumination that satisfies these conditions is said to be *critical* and can be produced by the two kinds of illuminating system described below.

4.1 Critical and Köhler illumination

The simplest arrangement for *critical illumination* is shown in Fig. 4.1. An image S' of the light source S is formed in the object plane of the microscope P by the substage condenser C. An iris diaphragm F, placed close to the light source, controls the area of the field that is illuminated, i.e. it acts as a field stop. Simultaneously the substage iris diaphragm A in the front focal

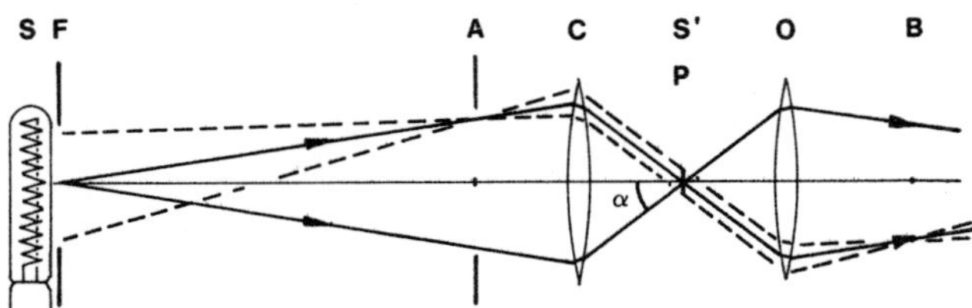

Figure 4.1 Principle of critical illumination. Images of the light source S and the field stop F are formed in the object plane, while an image of the aperture stop A is formed in the back focal plane of the objective B.

plane of the condenser controls the angle α of the cone of rays illuminating the object and entering the objective, i.e. it acts as an aperture stop. Rays of light coming from the plane of the substage diaphragm pass through the object plane as parallel pencils inclined to the optic axis, enter the objective O and are brought to focus in its back focal plane B.

The NA of the illuminating system must be as great as that of the objective if the maximum resolution of which the objective is capable is to be attained, hence α must be large. With such a simple illuminating system the light source needs to be large, say 25–50 mm diameter, when objects are to be examined with high-power objectives if maximum resolution is to be achieved. Thus in practice additional condensers are often included in the illuminating system to permit the use of a small source. Their inclusion also enables the source to be further away from the microscope stage, which permits the insertion of filters, etc., in the illuminating system.

This type of illumination has the disadvantage that any structure in the light source, e.g. filament coils in a tungsten lamp, is superimposed on the image. It is overcome either by throwing the light source slightly out of focus or by using a diffusing screen.

In *Köhler illumination*, light from a small source S is focused by a well-corrected lamp condenser C_L onto the iris diaphragm A that lies in the front focal plane of the substage condenser C_S (Fig. 4.2). Thus light from the plane of the diaphragm A passes through the object plane of the microscope P as pencils of parallel rays inclined to the optic axis. These rays enter the microscope objective O and are brought to focus in its back focal plane B, i.e. the substage condenser diaphragm acts as an aperture stop. Simultaneously the substage condenser forms an image of the back face of the lamp condenser and the iris diaphragm F in the object plane P, thus F acts as a field stop.

The system enables a small high-intensity light source to be used because the back surface of the lamp condenser acts as a uniformly illuminated secondary source of large diameter. It is less sensitive to slight variations in intensity distribution over the surface of the primary source than is critical

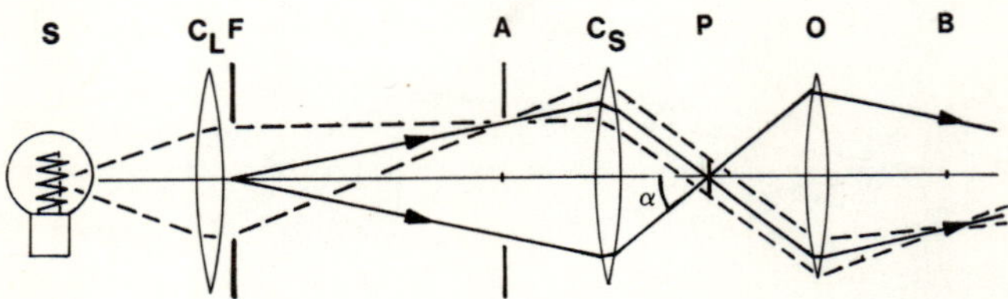

Figure 4.2 Principle of Köhler illumination. Images of the lamp condenser lens C_1 and field stop F are formed in the object plane P, while images of the light source S and aperture stop A are formed in the back focal plane of the objective B.

illumination, and this is particularly useful when the source is a coiled filament lamp. However, dirt and scratches on glass surfaces that are close to conjugate points may produce uneven illumination.

4.2 Effects of adjustment of the illuminating system

Irrespective of whether critical or Köhler illumination is used, the illuminating system must be correctly adjusted if we are to obtain optimum illumination. The aperture and field stops act independently; altering the setting of one does not affect the other, as we can demonstrate by a few simple experiments.

Focus the microscope on a specimen and, looking through the eyepiece, gradually close the field stop. A sharp image of the stop appears in the field of view and the illuminated area gradually becomes smaller, but its brightness remains unchanged. Next, with the eyepiece removed and the field stop fully closed, observe the back face of the objective lens, which approximates to the back focal plane of the objective, and then gradually open the field stop. We see that the whole of the back face is illuminated irrespective of the opening of the field stop but that the intensity of illumination increases as the stop opens. However, we see no image of the stop.

Observe the effect of changing from a low-power to a high-power eyepiece with the field stop partially closed. Increasing the eyepiece magnification produces a larger image of the stop, i.e. the diameter of the field of view decreases.

With the field stop open and the eyepiece in place, observe a specimen while gradually closing the aperture stop. The brightness of the field, but not its size, decreases and when the stop is closed well down diffraction fringes appear around the outlines of features in the microstructure. However, when we close the aperture stop with the eyepiece removed we observe that the size of the brightly illuminated disc on the back face of the objective progressively decreases, because the disc is an image of the aperture stop.

For optimum illumination we set the field stop so that only the field of view seen by the objective is illuminated, i.e. the field stop is opened until its image just disappears from the field of view, and the aperture stop so that about 80% of the area of the back face of the objective (about 90% of its diameter) is filled by the disc of light. If we open the field stop wider than this the glare in the system increases, while if we make the opening smaller we restrict the field of view. Similarly, if we open the aperture stop beyond its optimum setting it too contributes to glare, while if we close it below this setting it restricts the NA of the system.

Even though closing down the aperture stop reduces the brightness of the image it should *never* be used for this purpose, because it impairs the resolution of the image. Brightness should be reduced either by adjusting the brightness of the source of illumination, or by inserting a neutral filter in the illuminating system. Occasionally, when resolution is not critical, closing the aperture stop a little may be helpful in increasing contrast and depth and flatness of field, but the advantage is slight with good multicoated lenses.

If necessary, the adjustments of the stops should be carried out every time the objective is changed. They are quick and easy to perform and, if at first we take care to make them every time we change an objective, they soon become an essential part of the technique of using the microscope that we carry out instinctively.

4.3 Condensers and illuminators

Up to a total magnification of about × 25 no condenser is necessary, but at higher magnifications objectives cannot give their optimum performance unless the object is properly illuminated. The overall resolving power depends on the NAs of both the objective and the illuminating system. Provided that the NA of the objective exceeds that of the illuminating system, the resolution R is given by

$$R = 0.5\,\lambda/0.5\,(\text{objective NA} + \text{condenser NA}) \qquad (4.1)$$

Thus if a microscope is to give its optimum performance the NA of the condenser must equal that of the objective.

We can demonstrate the relative importance of the condenser and objective NAs on the resolving power of a microscope by using a transmitted-light microscope with a low-power objective and a well-dispersed powder specimen. With the stops in the illuminating system properly set, sharply focus the objective on a small group of particles and then observe the effect of gradually closing the aperture stop. The resolution becomes slightly impaired, the depth of field increases and eventually haloes form around particles. These are due to interference effects and they become very marked when the condenser NA is much less than that of the objective.

Reset the aperture stop, reduce the aperture of the objective by placing a disc of black paper with a tiny central pinhole over its back lens, and focus it on the specimen. We see that the resolution is grossly impaired, the depth of field has markedly increased and that closing the aperture stop reduces the brightness of the image but does not cause haloes to appear around the particles. Remove the disc of paper from the objective and estimate the size of the pinhole in the paper. Focus the microscope and close the aperture stop so that the disc of light at the back of the objective is about the same size as the pinhole. Observe that the resolution is much better than with the paper disc in position and that haloes are present around the particles. This shows clearly that the resolving power depends mainly on the size of the unobstructed NA of the objective.

Ideally an objective could be used as a condenser, indeed in the metallurgical microscope this is done because it is necessary to introduce the illumination behind the objective. However, for a transmitted light microscope a second objective would be mechanically inconvenient, and the corrections required for the lenses in the illuminating system are less stringent than those for objectives, because the illuminating sysem does not form an image that will be examined closely. Consequently, simpler, cheaper lenses can be used for condensers. In practice, a single condenser can be used for several transmitted light objectives, thus the microscope stand is fitted with a single substage condenser.

4.3.1 *Condensers for transmitted-light microscopes*

For use with medium- and high-power objectives substage condensers must be centrable, because it is essential that the optic axes of the microscope and illuminating system are aligned. Moreover, they must be focusable, because the position of the condenser is important in obtaining optimum illumination. Because a condenser has to provide illumination for a range of objectives from low power to high power, it must be able to deliver a beam of light in the object plane that is sufficiently broad to illuminate the relatively large fields viewed by low-power objectives, and a cone of rays whose angle is sufficiently large to match the NA of the highest-power objective. Moreover, field and aperture stops must be incorporated in the illuminating system so that the sizes of the illuminated field of view and aperture can be adjusted.

The simplest substage condenser is the two-lens Abbe condenser, which is cheap and has a long working distance. However, it suffers from both spherical and chromatic aberration, the former being the more objectionable. When optimum resolution is required this type of condenser is only suitable for use with lenses of low NA. Abbe condensers having three lenses are also made, and are better corrected for spherical aberration than two-lens condensers. Even they are unsuitable for use with objectives with NAs exceeding 0.6, owing to their spherical aberration. More complex aplanatic condensers that are better corrected for both spherical and chromatic aberration are suitable for use with objectives of higher NA.

Some condensers intended for use with high-power objectives have top lenses that can be swung out of the optical system. With the top lens in, the working distance is short and the size of field illuminated is small. This arrangement provides good illumination for high-power objectives, but it is not satisfactory for low-power objectives. When the top lens is removed from the optical system the size of field illuminated is increased, which enables the condenser to be used with low-power objectives, but the substage diaphragm is made ineffective.

When using oil immersion objectives the top lens of the condenser must be oiled to the underside of the slide, as well as the objective to the specimen. If this is not done the optimum resolution of the objective cannot be attained.

4.4 Illumination of opaque specimens

Because the working distances of most objectives are short, illumination can rarely be introduced between objective and specimen. Thus, oblique illumination can only be used at low magnifications up to about $\times 50$. Under oblique illumination surfaces perpendicular to the optic axis of the microscope appear dark, because the light is not reflected into the objective, while features on the surface that reflect or diffract light into the objective, such as steps and grain boundaries, appear bright (Fig. 4.3).

The most useful kind of illumination is normal incident or vertical illumination, which requires a reflector to be introduced into the optical system of the microscope behind the objective. The light is reflected down the optic axis of the microscope usually by a thin glass slip reflector inclined at an angle of 45° to the optic axis (Fig. 4.4*a*). For maximum efficiency the slip and objective lenses must be coated, but even so only about 20% of the light is available for image formation. A serious problem with the arrangement is that when polarized light is used the plane of polarization is markedly rotated as it passes through the slip. This difficulty is overcome by the Smith illuminator (Fig. 4.4*b*), in which the slip is inclined at about 67.5° to the optic axis. More intense illumination can be obtained by the use of a prism or mirror that partially obstructs the microscope tube (Fig. 4.4*c*). These act as unsymmetrical aperture stops and produce less perfect images

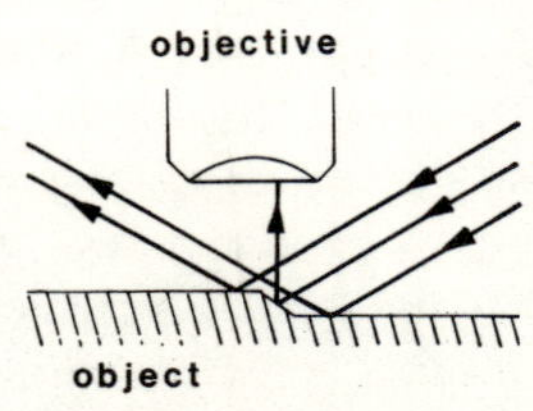

Figure 4.3 Oblique illumination of an opaque specimen.

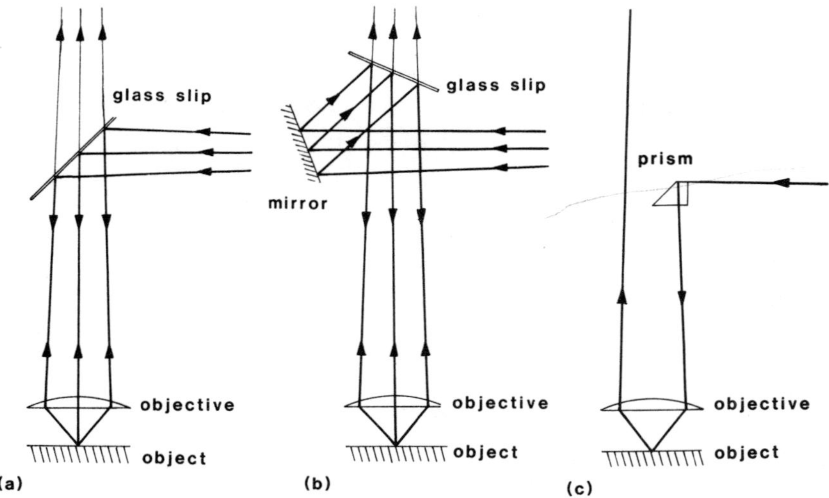

Figure 4.4 Forms of incident light illuminators for metallurgical microscopes. (*a*) Inclined slip illuminator; part of the light reflected by the specimen passes through the slip to the eyepiece; (*b*) the Smith illuminator; (*c*) prism illuminator.

than do inclined slip illuminators. For this reason and because excellent high intensity light sources are available they are rarely used in modern microscopes.

Because the light is introduced into the optical system behind the objective, it is used as its own condenser. Since it is impractical to have an aperture stop in the back focal plane of the objective, the illuminating system usually incorporates an aperture stop whose image is focused in the back focal plane of the objective by an auxiliary lens and a field stop whose image is focused in the object plane by the objective.

4.4.1 *Dark-field illumination*

For dark-field illumination, which is a special case of oblique illumination, the specimen is illuminated with a hollow cone of light that does not enter the objective. The image is formed by rays of light that are diffracted or reflected into the objective from detail in the specimen. Consequently, detail appears bright while the remainder appears dark (Fig. 4.5). The intensity of illumination is low but this helps to give sharper contrast than bright-field illumination. Moreover, the colour of inclusions and precipitates in opaque specimens may show more clearly. Thus the technique sometimes is useful for increasing the visibility of detail; no increase in resolving power occurs, indeed resolution may be diminished. Interpretation of the images is more difficult than that of bright-field images owing to interference effects.

Figure 4.5 Dark field photomicrograph of pearlite, NA = 0.55, magnification × 500. Compare this micrograph with the bright field micrograph of pearlite in the same specimen, Fig. 3.4*b*.

Conventional illuminating systems for transmitted-light microscopy are only satisfactory for dark-field illumination with lenses of fairly low NA, < 0.5. In these cases a central patch stop of appropriate size mounted beneath the substage condenser may be used to produce the hollow cone of light. However, for the majority of objectives of moderate and high NA special catoptric (i.e. mirror) condensers need to be used.

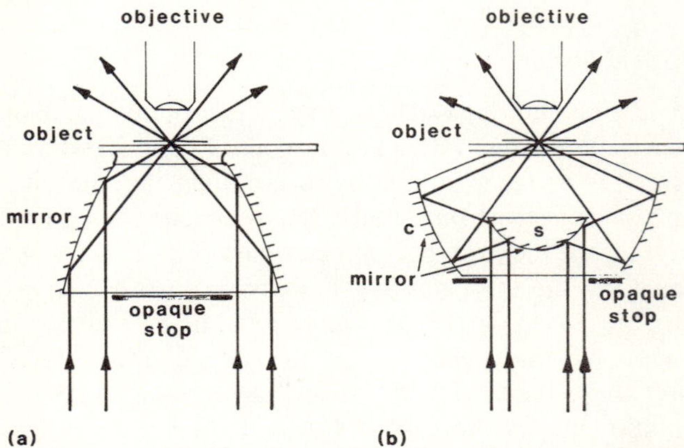

Figure 4.6 Principles of dark field illuminators for transmitted light microscopes. (*a*) Paraboloid catoptric condenser, (*b*) cardioid catoptric condenser.

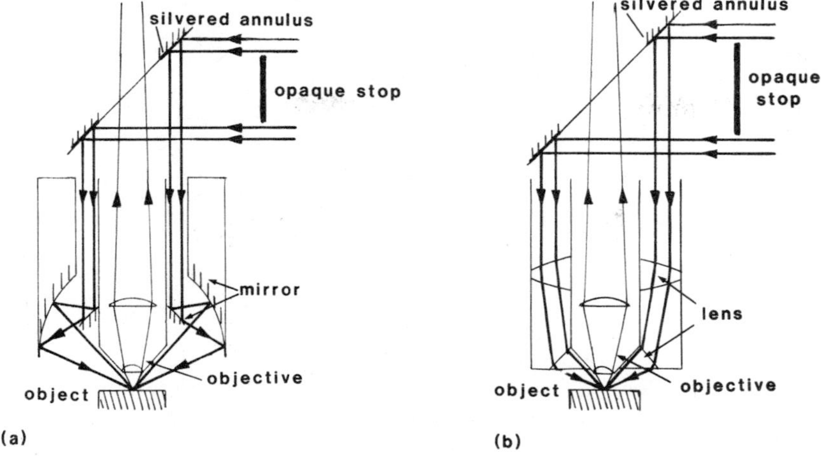

Figure 4.7 Principles of dark field illuminators for metallurgical microscopes. (*a*) Cardioid catoptric condenser, (*b*) dioptric condenser.

There are two types of catoptric condenser, the paraboloid and the cardioid. In the former (Fig. 4.6*a*) an annular collimated beam of light falls on a parabolic mirror that focuses it in the object plane, whereas in the latter (Fig. 4.6*b*) an annular collimated beam of light is reflected between a spherical mirror *S* and a cardioid mirror *C* to focus it in the object plane. Better correction for spherical aberration is possible with the cardioid condenser, hence it is the preferred type. With objectives of high NA, including dry ones, it is necessary to oil the condenser to the slide in order to obtain a hollow cone of rays of sufficiently wide angle. Accurate centring and focusing of the condensers is necessary.

In reflected-light microscopy, dark field illumination may be produced either by a cardioid catoptric condenser or by a dioptric (i.e. lens) condenser in the centre of which the objective is mounted (Fig. 4.7). An annular beam of light, produced by inserting a patch stop in the illuminating system, is reflected by an aluminized annular reflector into the condenser (Fig. 4.7). In practice each objective often requires its own illuminator, hence some manufacturers supply objectives with built-in dark-field illuminators.

Dark-ground illumination should be distinguished from annular illumination. In the latter oblique lighting of radial symmetry that *enters* the objective illuminates the specimen. Such lighting sometimes increases the visibility of details in the transmitted-light microscope, but it does not improve the resolution. Indeed, because the image is formed by interference of diffracted rays the image formed may be less perfect than when a solid cone of light is used.

5 Polarized light microscopy

The waves in a beam of *ordinary* light vibrate in all directions transverse to the direction of propagation. However, under certain conditions some directions of vibration are eliminated or rotated so that the vibrations occur either in a single plane or in a pair of planes: in the former case we say that the light is plane polarized and in the latter, in general, is elliptically polarized. In the polarizing microscope we make use of the effects of crystals and other constituents on plane polarized light to study and characterize the specimen.

5.1 Polarized light

5.1.1 *Polarization by reflection and refraction*

When a beam of ordinary light falls onto the surface of a transparent isotropic body part of the light is transmitted and part is reflected. The two beams are partially polarized in mutually perpendicular planes (Fig. 5.1),

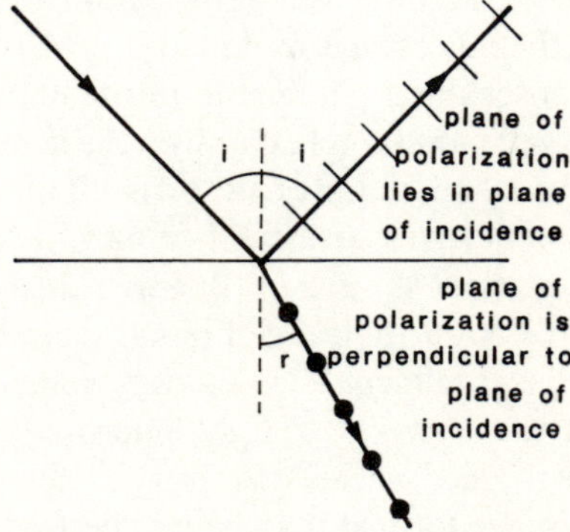

Figure 5.1 The reflected and refracted beams formed when an unpolarized beam of light falls on a transparent surface are partially polarized in mutually perpendicular planes.

and the plane of polarization of the *reflected* beam lies in the plane of incidence. Polarization is incomplete but reaches a maximum when the two beams are perpendicular, when the reflected beam, but not the refracted beam, is completely polarized. The critical angle of incidence for polarization, the polarization angle θ_c, is given by Brewster's Law

$$\tan\theta_c = n_1/n_2 \tag{5.1}$$

where n_1 and n_2 are the refractive indices of the media above and below the reflecting surface. For an air-crown glass ($n = 1.52$) interface θ_c is $\sim 57°$. Incidentally, this law enables the R.I.s of opaque materials to be determined by measuring θ_c.

When a polarized beam of light falls onto a transparent surface its behaviour depends on the orientation of the plane of polarization with respect to the surface. If the plane is transverse to the surface the beam enters the material, but if it is perpendicular the beam is reflected; in both cases the planes of polarization are unchanged.

5.1.2 *Elliptically and circularly polarized light*

When two coherent rays that differ in phase and are polarized in different planes follows the same path their linear vibrations combine to produce a vibration that is generally elliptical, i.e. a point vibrating under the influence of the waves would follow an elliptical path. Such light is said to be elliptically polarized. Two special cases occur when the planes are mutually perpendicular and the amplitudes of the waves equal. When the phase difference between the waves is a quarter wavelength the resultant vibration is circular, thus the light is circularly polarized. Moreover, when the phase difference is a half wavelength the resultant vibration is linear, i.e. the light is plane polarized.

Elliptically polarized light is produced when parallel overlapping ordinary and extraordinary rays emerge from a birefringent crystal, as described in the following section. It is also produced when plane polarized light is reflected at an anisotropic surface.

5.1.3 *The effect of crystals on polarized light*

Crystals possessing a cubic structure and non-crystalline materials, e.g. glass, are isotropic, i.e. their properties are the same in all directions. Within them light vibrates with equal ease in all directions perpendicular to the direction of propagation, consequently they have no effect on rays of polarized light that pass through them.

However, when a ray of polarized light enters an anisotropic crystal, other than along an optic axis, it is reorganized into two rays that vibrate in fixed planes at right angles to each other, and that follow different paths in the crystal. One of the rays travels with the same velocity in all directions and obeys Snell's law, hence it is called the ordinary or O-ray. The other, called the extraordinary or E-ray, travels with a velocity that is direction-dependent and does not obey Snell's Law. Such a crystal is said to be double refracting or birefringent and the maximum difference between the R.I.s of the two rays is called the birefrigence of the crystal.

Provided that they overlap, when the two sets of rays emerge from the crystal they recombine forming an elliptically polarized beam, the ordinary ray derived from one incident ray combining with an extraordinary ray from another. However, when the incident beam is unpolarized no effect is seen because all the planes of vibration of the waves in the incident beam are affected to the same extent. Thus the distribution of vibration planes in the transmitted beam reproduces that in the incident beam.

5.1.4 *Polarizers or polars*

When we view a point through a suitably oriented birefringent crystal, e.g. calcite, we see two images of it because the two rays follow different paths. This behaviour can be used to produce a beam of plane polarized light for microscopy by isolating one of the rays. Traditionally this was done using a double calcite prism of special geometry, called a Nicol prism, that transmitted the E-ray but eliminated the O-ray by making it diverge strongly so that it did not enter the microscope. Nowadays, polarizing filters that separate the two rays and completely absorb only one of them are used to produce plane polarized light for microscopy. They consist either of a film of oriented dichroic crystals in an isotropic matrix, or a plastic film in which polymer molecules have been oriented by stretching, to which has been added an absorbent dye that absorbs completely only one of the rays.

Such devices are called polarizers or polars. Modern polarizing filters produce as good or better images than Nicol prisms, they permit the use of much broader beams and are much cheaper.

5.2. The polarizing microscope

A polarizing microscope differs from an ordinary transmitted-light or incident-light microscope in having polarizing filters, the polarizer and the analyser, fitted in the illuminating system and the microscope tube. At least one of the filters must be mounted in a rotatable, graduated mount. Nevertheless usually they are set with their planes of polarization perpendicular to each other and parallel to the cross wires in the eyepiece. One or both can be removed from the optical path as required, thus enabling the specimen to be viewed in ordinary or plane polarized light, or between crossed polars. A slot for the insertion of accessory plates etc., at 45° to the planes of polarization of the crossed polars, is provided in the microscope tube behind the objective. Also there is provided a Bertrand lens for the examination through the eyepiece of interference figures in the back focal plane of the objective. When required it is inserted in the optical path of the microscope just beneath the eyepiece. The stage must be able to rotate and the rotation must be graduated, and the stage and objectives must be mutually centrable. The objectives must be strain-free, because strain in them causes birefringence.

5.2.1 *Crystals and the transmitted-light polarizing microscope*

Look through a transmitted-light microscope with the polarizer and analyser in the optical system and no specimen on the stage, and rotate one of them. Observe that the light is extinguished twice in a complete revolution. Extinction occurs when the planes of vibration of the polars are perpendicular to each other and parallel to the crosswires in the eyepiece.

Next place birefringent crystal on the stage between *crossed* polars and *rotate* the *crystal*. The light is now extinguished four times in each revolution (Fig. 5.2). In this case, extinction occurs when the planes of polarization of the crystal become parallel to those of the polars. In

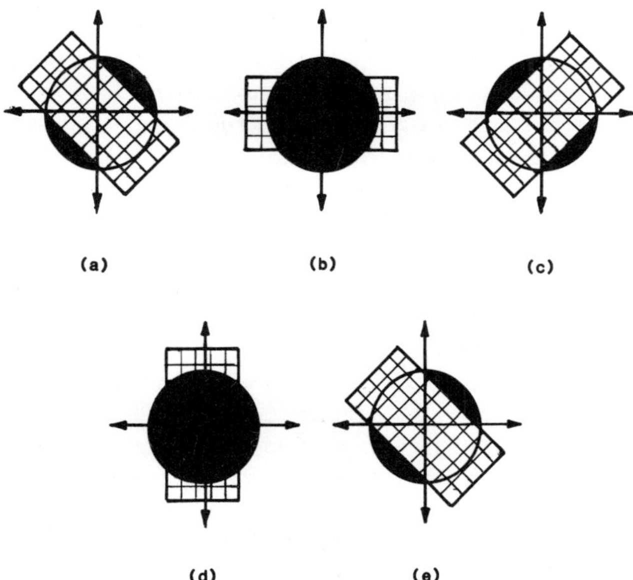

Figure 5.2 The effect on light transmission of rotating a birefringent crystal between crossed polars.

intermediate positions light passes the analyser becoming gradually more intense until the 45° position is reached.

5.3 Interference colours

When white light is used, anisotropic crystals may appear coloured when viewed between crossed polars. This occurs owing to interference effects between rays emerging from the analyser.

First consider the passage of a beam of plane polarized *monochromatic* light through a birefringent crystal not in an extinction position. On entering the crystal the beam is resolved into O- and E-rays that travel with different velocities through the crystal. When they emerge from the crystal the phase of one set of rays has been retarded with respect to the other and, in general, O- and E-rays following the same path combine to produce elliptically polarized light. When this beam falls on the analyser the vibration directions of its components, which are parallel to those of the O- and E-rays emerging from the crystal, do not coincide with the vibration direction of the analyser. Consequently each of the components of the beam is further resolved into O- and E- components in the analyser, whose vibration directions are respectively parallel and perpendicular to the direction of vibration of the analyser. Those whose vibration directions are perpendicular to that of the analyser are absorbed, while those whose directions are parallel are transmitted. The two sets of rays transmitted by the analyser either combine or interfere, the effect produced depending on the phase difference between the O- and E-rays leaving the crystal and their amplitudes. In the polarizing microscope complete extinction occurs when the phase difference is a *whole* wavelength or multiple of it, owing to the effect of the analyser. When this condition is satisfied the light is extinguished for all positions of the crystal, and not only when the vibration directions in the crystal are parallel to those of the polars.

The retardation of one wave behind the other depends on the difference in wave velocities and the thickness of the crystal. Because the difference in wave velocities is a function of wavelength, it follows that for a given thickness of crystal plate total extinction will occur for, at most, a few wavelengths. Consequently when the crystal plate is illuminated with white light, between the four extinction positions the crystal shows interference colours that are subtractive colours, i.e. white light minus the wavelengths that have been extinguished. The wavelength extinguished increases with the thickness of the crystal plate and in thicker plates the condition for extinction may be satisfied for several wavelengths. This gives rise to interference colours that are grouped in the seven orders of Newton's Scale. The colours seen depend on the birefringence of the crystal, its thickness, and the orientation of the section relative to the crystal axes.

Sometimes birefringence is combined with the selective absorption of light in one vibration plane; this is called *pleochroism.* The effect is seen when the crystal is rotated in plane polarized light with the analyser withdrawn. In white light the colour of the crystal changes with orientation, e.g. biotite changes from pale yellow to dark brown. Dichroism is an extreme form of pleochroism in which one of the rays is completely absorbed. Under the same conditions crystals that exhibit high birefringence but no selective absorption 'twinkle' owing to marked variations in their visibility.

5.4 Polarizing microscope accessories

The accessories are used to help interpret the effects produced by polarized light. They consist of thin plates or wedges of birefringent materials cut in specific ways relative to the crystallographic axes of the materials. They are mounted in slides with one of their vibration directions parallel to the long edge of the slide, so that they can be inserted behind the objective in a slot at 45° to the vibration directions of the polars (Fig. 5.3). One or both of the vibration directions of the waves are marked on the slides.

5.4.1 *The quarter-wave plate*

A quarter-wave plate is made from a flake of mica that is cleaved to such a thickness that the O- and E-rays emerge a quarter of a wavelength out of phase, corresponding to a pale grey interference colour. With transparent crystals we use the plate by rotating the crystal to 45° from its extinction position, noting its interference colour and then inserting the plate and

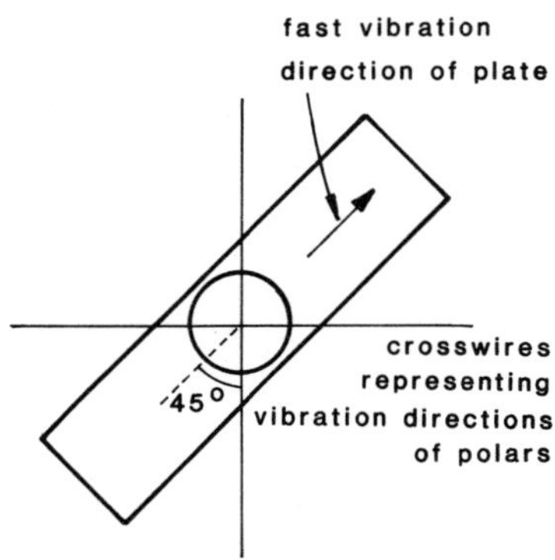

Figure 5.3 Accessory plates are inserted into the optical path of the microscope so that their vibration directions are at 45° to those of the polars.

noting the new interference colour and whether it is of lower or higher order than the first colour. The order of the interference colour increases when like vibrations are superimposed and vice versa, thus the fast and slow vibration directions of the crystal are identified. The quarter-wave plate is especially useful for examining sections showing bright (second- to fourth-order) interference colours, because the interference colours are moved only a short distance along the scale, reducing the risk of confusion compared with use of the whole-wave plate.

5.4.2 *The sensitive-tint or whole-wave plate*

Sensitive-tint plates are slices of birefringent material, usually gypsum, mica or quartz, cut or cleaved parallel to the optic axis of the crystal to such a thickness that the O- and E-rays for green light (540 nm) are out of phase by exactly one wavelength. When the microscope is focused on a plane *isotropic* specimen between crossed polars and the sensitive-tint plate interposed in the 45° position the plate introduces no ellipticity into the plane polarized green light because the emergent O- and E-rays are in phase and combine to form a plane polarized beam. Thus the analyser extinguishes the green light but permits all other wavelengths to pass to some extent, depending on the degree of ellipticity for each wavelength. Hence the light passing through the analyser is white minus green and appears magenta ('sensitive purple') in colour. This colour divides the first and second orders of Newton's Scale. The plate may be used in same way as the quarter-wave plate to determine the fast and slow vibration directions of sections that show low-order interference colours. Also it is used to introduce colour contrast in polarized light images, because colour often enhances visual contrast, especially in metallographic images.

5.4.3 *The quartz wedge*

A quartz wedge works in the same way as the plates, but has the advantage that it covers several order of colour on Newton's Scale. With the polars crossed and the crystal in the 45° position, the quartz wedge is slowly inserted and the position at which complete extinction occurs is noted. This only occurs if the order of the colour decreases; if it increases the crystal is turned through 90° to cause the order to decrease. This is best ascertained by observing the back focal plane of the objective, in which a narrow black line appears when extinction occurs. Since compensation only occurs when unlike vibrations in the crystal and wedge are parallel, it is a guide to the optic sign of the crystal.* It is also used to determine the true colour value of higher-order interference colours.

* A uniaxial crystal is optically positive when the O-ray is faster than the E-ray.

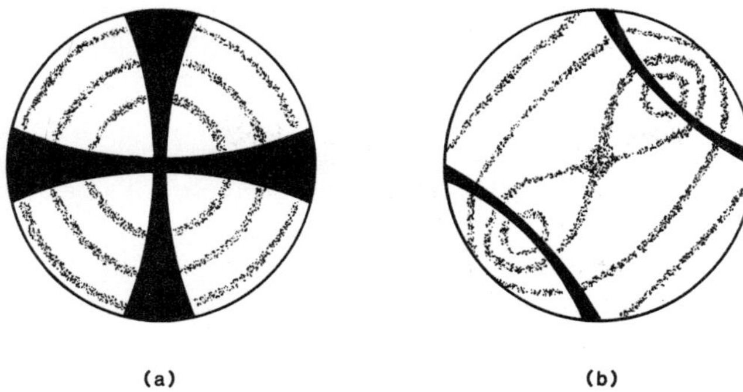

Figure 5.4 Examples of interference figures. (*a*) Basal uniaxial figure, (*b*) a biaxial figure.

5.5 Interference figures

The optic sign of an anisotropic crystal is determined by examining the interference figures that are formed in the back focal plane of a high-power objective (e.g. 4 mm focal length) when the crystal is examined between crossed polars using *convergent* light, i.e. conoscopic illumination. The interference figure is viewed either by removing the eyepiece or by inserting the Bertrand lens into the microscope tube beneath the eyepiece. All sections of birefringent materials produce interference figures, but the information that can be deduced from the figure depends on the section orientation. The most useful figures are those in which the points of emergence of the optic axes can be seen.

The interference figures consist of black or shadowy isogyres and coloured bands that are elliptical or circular in shape, whose colours and spacings depend on the birefringence of the crystal (Fig. 5.4). The isogyres may be in either the form of a cross or of hyperbolae. Determination of the optic sign from the figure is achieved by the use of accessory plates or the quartz wedge.

5.6 Crystals and the incident-light polarizing microscope

When plane polarized light falls normally on to an opaque *isotropic* surface it is reflected without the state of polarization being altered. However, if the angle of incidence differs from this the reflected light is elliptically polarized and cannot be completely extinguished by the analyser unless the angle of incidence of the light lies in the vibration plane of either of the polars. Thus it is advantageous to close down the aperture stop to eliminate oblique boundary rays. If the back focal plane of the objective is examined a dark cross is seen when the polars are crossed.

On the other hand, when a plane polarized beam of light falls on an anisotropic crystal surface it is resolved into two plane polarized components that vibrate in directions perpendicular to each other. The positions of these directions depend on the crystal orientation. The reflecting powers of the crystals in the two directions usually differ and there is a phase difference between the components. However, if the *c*-axis of a uniaxial crystal is perpendicular to the surface it behaves in the same manner as an isotropic crystal, which is analogous to its behaviour in transmitted light. With *metals* the main effect is the difference in reflecting power of the two components; in most cases the phase difference is small. This causes the plane of vibration of the polarized light to rotate when reflected, permitting a component to pass the analyser and the surface appears more or less bright, depending on the intensity of the component passing the analyser. In the absence of a phase difference between the components, when the vibration directions in the polars correspond to those of the vibration directions of the crystal surface no rotation of the plane of polarization occurs, and the reflected light is extinguished. Thus when the crystal is rotated the light is extinguished four times in each revolution. However, if in addition to the difference in reflecting power there is a phase difference between the components the reflected light is elliptically polarized and complete extinction is not possible.

5.6.1 *Reflection pleochroism*

When we examine isotropic materials in plane polarized light with the analyser withdrawn we observe no brightness changes when they are rotated, but we do observe brightness changes with anisotropic materials. Maxima or minima in brightness occur when the vibration directions of the crystal surface are parallel to the plane of vibration of the incident light; usually the effect is small but in strongly anisotropic crystals it is easily seen.

If the reflecting power of the material alters with wavelength the reflected light shows a characteristic tint. Moreover, in anisotropic materials the characteristic tint may be different for the two vibration directions, and can also be seen between crossed polars. The effect is a characteristic feature of some constituents and inclusions in alloys and is an aid to identifying them, e.g. unetched $CuAl_2$ in aluminium alloys shows orange to green-blue colours when rotated between crossed polars.

5.6.2 *Etched specimens*

The change in the state of polarization of the reflected beam depends on the angle between the surface and the incident beam. Consequently when

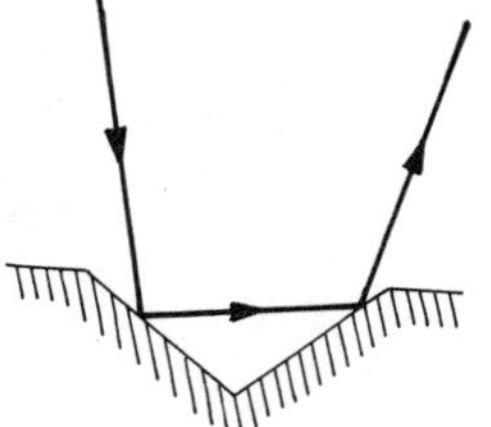

Figure 5.5 Reflection of a light ray within an etch pit or groove.

examined between crossed polars inclined surfaces show four minima and maxima in brightness when they are turned through 360°. Both isotropic and anisotropic crystals develop geometrical faceted etch pits or grooves (strings of etch pits) in the surface when etched with suitable reagents. Reflection of light rays within a pit (Fig. 5.5), introduces ellipticity in the reflected ray. Because all the etch pits or grooves in a grain are identically orientated, the reflections from all the pits in a grain are identical, but different from those in neighbouring grains. Thus between crossed polars a polycrystalline specimen will show grain contrast. Moreover, when the specimen is rotated each grain will extinguish four times in a complete revolution. The visual contrast is enhanced by use of a sensitive-tint plate, which causes grains of different orientations to appear in different colours; this is a very effective method of, for example, showing up preferred orientations and of identifying annealing twins in metals.

5.7 Uses of polarized light in the microscopy of materials

5.7.1 *Transparent materials*

The major application of polarized light is in the examination of thin sections of transparent materials. Polarized-light microscopy of these materials is of comparable importance to bright-field microscopy in metallography. The use of polarized light yields much more information than can be obtained from bright-field illumination and in its more sophisticated applications is capable of yielding accurate quantitative information about the optical properties of the constituents present. Such information can be used to positively identify constituents, since the optical properties of numerous minerals, etc. are tabulated; it is also useful in constitutional studies. It enables isotropic and anisotropic crystals to be quickly and unambiguously distinguished, and is an effective method of introducing contrast between grains, often revealing detail and morphology absent or not clearly visible in bright-field illumination (Figs. 5.6, 5.7). The technique is used in the examination of ores, ceramic raw materials

Figure 5.6 Concrete roof tile. Crossed polars (× 40). Rounded quartz grains (various shades) are embedded in a cement matrix (black). The polarized light shows that the larger quartz grains are polycrystalline.

Figure 5.7 Spherulites in a thin film of polypropylene grown from the melt at ~ 130°C. Crossed polars (× 125). In bright field illumination the spherulite boundaries are only faintly visible.

and products, refractories, slags, glasses, enamels, cements, abrasives, organic polymers, liquid crystals and woods.

Isotropic transparent materials, both crystalline and amorphous, become birefringent when stressed. For this reason optical components of polarizing microscopes must be stress-free. Stress birefringence is used to investigate internal stresses in both transparent isotropic and anisotropic materials. An example of the technique is the revelation of slip planes as bright bands in deformed crystals of lithium fluoride when viewed between crossed polars.

5.7.2 *Opaque materials*

Examination of the freshly polished surfaces of anisotropic metals and alloys and ceramics between crossed polars produces grain contrast in the image and clearly shows twinning and preferred orientation, provided that the surface is properly prepared, and is free from scratches, surface relief and cold work. Often it is necessary to electropolish metallic surfaces, because mechanical polishing distorts the surface layer: nonetheless some inves-

Figure 5.8 Spheroidal graphite cast iron. Etched in 2% nital (× 500). The spherulitic structure of the unetched graphite nodules is revealed by the polarized light; the dark crosses in the spherulites are parallel to the vibration directions of the polars and arise in a similar manner to isogyres in interference figures. The nodules are surrounded by envelopes of ferrite (black), whose grain boundaries appear light, and these are embedded in a pearlite matrix, whose structure is shown by the polarized light.

tigators have successfully used mechanical polishing. The technique is especially useful when the material cannot be satisfactorily etched, e.g. graphite spheroids in cast iron (Fig. 5.8).

Isotropic metals also can be studied using polarized light if anisotropic epitaxial films can be formed of their surfaces. They can be formed on some metals by anodic oxidation, or by the controlled deposition of thin oxide or sulphide layers. Moreover, etching of polished surfaces to produce etch pits or grooves also renders isotropic materials responsive to polarized light, e.g. pearlite (Fig. 5.8). In some cases, e.g. anodizing of aluminium, it is not clear whether it is the anisotropic film or the pitting that causes the response to polarized light.

Intermediate phases and non-metallic inclusions in metals are often anisotropic and respond to polarized light. Thus some intermediate phases in aluminium alloys respond to polarized light by exhibiting characteristic polarization colours. Inclusions may be either transparent or opaque. Transparent inclusions reflect light both from the surface and internally, and in ordinary light their appearance depends on the contribution from both sources. However, between crossed polars the reflection from the surface is extinguished by the analyser, but that from the internal reflections is not, so that the inclusions show their transmission colours, e.g. cuprous oxide appears a blue-grey colour in ordinary light but is a bright ruby-red colour between crossed polars, which distinguishes it from cuprous sulphide whose colour remains blue-grey in both kinds of illumination. Moreover, hemispherical transparent inclusions often show concentric dark rings and a dark cross, due to reflections from the regular bottom of the inclusion. Opaque phases and inclusions if isotropic appear dark, but if anisotropic they may reveal their anisotropy, also they may exhibit reflection pleochroism. The behaviour of many non-metallic inclusions in polarized light is characteristic and may be used to identify them.

The polarizing microscope can also be used to study both opaque and transparent ferromagnetic materials, owing to the Kerr and Faraday effects respectively. When a specimen that is illuminated with plane polarized light is magnetized, so that there is a component of the specimen's magnetization along the optic axis of the microscope, the plane of polarization is rotated. This effect is useful in the study of magnetic domains.

6 Opaque stop and phase contrast microscopy

Both techniques employ a similar method of illuminating the specimen. Usually the experimental arrangement enables both techniques to be employed using the same simple equipment, hence they are considered together. The equipment required is relatively inexpensive and thus extends the capabilities of the microscope cheaply.

6.1 Opaque stop microscopy

Opaque stop microscopy, or sensitive dark-ground illumination, is essentially a reflected-light technique, which can be achieved by simple modification of the metallurgical microscope. An annular diaphragm, i.e. an opaque disc with an annular opening, is inserted in the illuminating system and aligned with an annular opaque stop in the microscope tube, behind the inclined glass slip illuminator* (Fig. 6.1). In practice, the central disc in the annular diaphragm fills about the central third of the area of the back lens of the objective, and the size of the annulus in the diaphragm is sufficiently large to allow full use of the NA of the objective.

Because of the annular diaphragm the specimen is illuminated by a hollow cone of light. We set up the arrangement using a highly polished, perfectly flat specimen with its surface perpendicular to the optic axis of the microscope. After focusing on the specimen, we bring the image of the annular diaphragm to focus in the plane of the opaque stop with the aid of an auxiliary microscope used in place of the eyepiece. Then we adjust the position of the annular diaphragm so that its image coincides exactly in size

* Ideally the opaque stop should be in the back focal plane of the objective, but in practice its position is not critical. In this case it is advantageous to site it behind the inclined slip illuminator. Nonetheless the distance between the back focal plane of the objective and the opaque stop should be as short as possible.

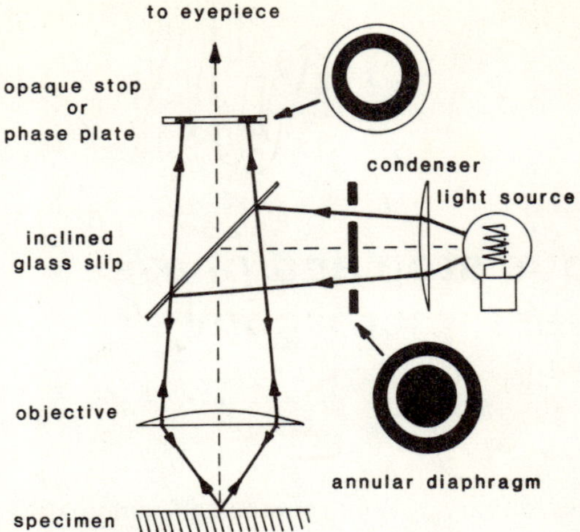

Figure 6.1 Schematic arrangement of the opaque stop and phase contrast microscope for use with reflected light.

and position with the annular opaque stop, thus eliminating the direct beam.

Suppose that we replace the well-polished specimen by one in whose field there are two surfaces inclined at a small angle ϕ, one of which is perpendicular to the optic axis of the microscope (Fig. 6.2). Images of the two surfaces are formed that are slightly displaced with respect to each other by an angle 2ϕ. The rays of light from the surface that is perpendicular to the optic axis are obscured by the opaque stop, while those from the inclined surface miss the stop, thus changing the illumination from true dark-ground to Schlieren arrangement. In general, geometrical deviations, either by reflection or refraction, produce image contrast because light from some areas is interrupted by the opaque stop while that from others is not. (Also, diffracted rays miss the opaque stop and contribute to image formation.) This makes visible slopes and tilts that cannot be detected in bright-field illumination. It has been shown that at magnifications greater than about $\times 50$ the gain in sensitivity is about thirtyfold, and is similar to that for dark-field or conventional oblique illumination.

When a sloping region deflects the direct beam it is no longer absorbed by

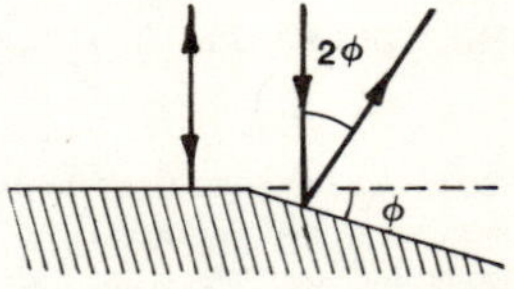

Figure 6.2 A small surface tilt through angle ϕ displaces the images of the annular diaphragm through angle 2 ϕ.

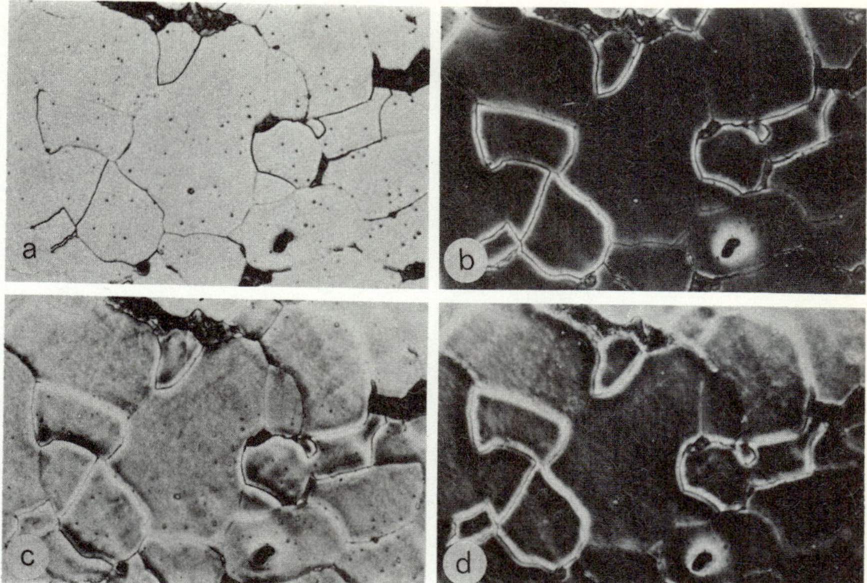

Figure 6.3 Mild steel. Lightly etched in 2% nital (× 250). (*a*) Bright field illumination; (*b*) opaque stop illumination; (*c*) positive phase contrast illumination; (*d*) negative phase contrast illumination. Note the halo effect at grain boundaries in (*b*), (*c*), and (*d*).

the opaque stop. However, by moving the annular diaphragm true dark-ground illumination can be restored. After calibration, the amount of movement can be used to measure the inclination of the sloping region; a sensitivity of measurement of about 1° is obtained at a magnification of about × 200 using an objective of 0.45 NA.

As well as revealing surface tilts as bright images on a dark field, the technique also reveals fine surface detail in etched specimens, including scratches, as bright features on a dark background (Fig. 6.3).

6.2 Phase contrast microscopy

Some specimens are unable to produce sufficient contrast to enable the detail present in them to be visible in bright-field illumination. Nonetheless, slight phase changes are present in the transmitted or reflected beams, and phase contrast microscopy, a special case of interference microscopy, manipulates these differences to make the detail visible.

There are two kinds of specimen with which phase contrast microscopy may be useful. Firstly, transparent specimens that have regions that differ slightly in optical density, or in which there are slight differences in optical density within a region. Secondly, opaque reflecting specimens that show small variations in surface level.

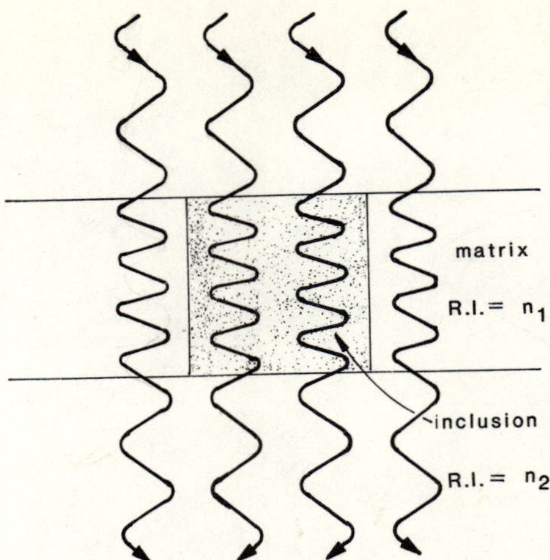

Figure 6.4 Illustrating a transparent specimen with an inclusion whose R.I. differs from that of the matrix, $n_2 > n_1$, illuminated by coherent light. The rays passing through the inclusion are retarded by half a wavelength relative to those passing through the matrix. The two sets of rays interfere destructively when they emerge from the specimen.

Figure 6.4 shows a section through a transparent, non-absorbing specimen that consists of a matrix containing an inclusion of higher R.I., illuminated by a beam of coherent light. As the beam passes through the specimen, the phase of the rays passing through the inclusion lags behind that of the rays passing through the matrix. When the two sets of rays, which are of equal intensity, emerge from this specimen they are out of phase by exactly half a wavelength and they interfere destructively causing the inclusion to appear dark, i.e. contrast is produced. In this extreme case the effect can be observed under bright-field conditions if the aperture stop is closed sufficiently to reduce interference by other diffracted rays. However, in specimens of normal thickness the phase shift produced is usually between an eighth and a quarter of a wavelength and a phase plate that either accelerates or retards the phase of the direct beam by a quarter of a wavelength is used to increase the degree of interference between the beams and thus increase contrast.

In the case of a reflecting specimen coherent rays reflected from the surface of the specimen will not be exactly in phase with those reflected from, e.g., the bottom of a shallow pit or the top of a slightly raised area (Fig. 6.5). Other things being equal the maximum interference occurs when the path difference between the rays from the surface and the feature is half a wavelength.

We can understand the operation of the phase contrast microscope by considering the amplitude vectors OA and OB of the light waves from the

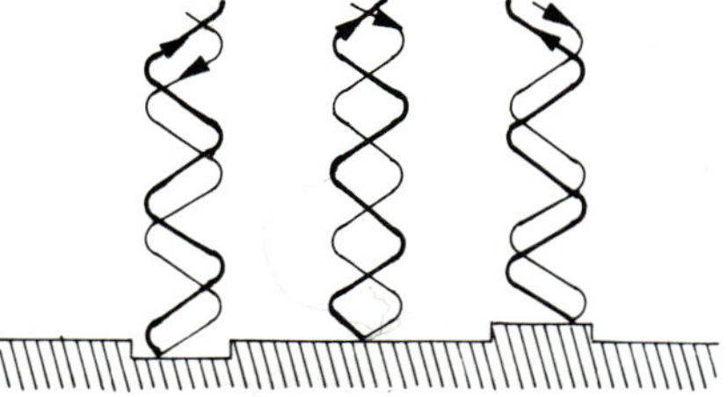

Figure 6.5 Phase differences between reflected rays are produced by differences in surface level.

reference constituent or surface and the feature (Fig. 6.6*a* and *b*). Both waves are of the same amplitude, i.e. $|OA| = |OB|$, but OB lags in phase by an angle ϕ. Consequently we cannot perceive the feature because the intensities of both waves, which are proportional to the squares of their amplitudes, are equal. To achieve amplitude contrast OB must be made smaller or larger than OA. The vector OB can be resolved into the two components OA and AB, i.e. $\mathbf{OB} = \mathbf{OA} + \mathbf{AB}$, the angle between which is ψ (Fig. 6.6*c*), and when ϕ is small $\psi \simeq 90°$, corresponding to a phase difference

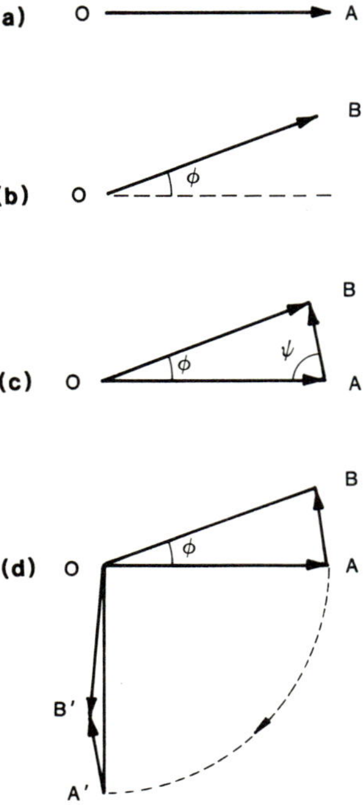

Figure 6.6 Vector diagram of image formation in phase contrast microscopy.

between OA and AB of about a quarter wavelength. OA represents the direct beam in the interference image that gives a uniform light distribution without detail. AB represents higher-order diffracted beams that produce image detail by interfering with OA, but the detail cannot be perceived by the eye because contrast is lacking, owing to AB being almost perpendicular to OA. If we introduce in the back focal plane of the objective a phase plate that advances the direct beam by 90° (a quarter wavelength), i.e. shortens the optical path of the beam, OA is rotated through 90° in a clockwise direction to OA', Fig. 6.6*d*. Since $\mathbf{A'B'} = \mathbf{AB}$ and $\psi \simeq 90°$, $A'B'$ now interferes with OA' destructively and the resultant vector is OB', which is less than OA' and is approximately equal to $OA' - A'B'$. Consequently the feature appears darker than the reference constituent or surface. In practice $A'B'$ is much smaller than OA' so that the darkening of the feature, and therefore the contrast, is slight. To enhance the contrast OA' is made comparable in magnitude to $A'B'$ by partially absorbing the direct beam coming from the specimen but not the diffracted beams. The effect produced by advancing the direct beam by 90° is described as positive phase contrast. Conversely, when the direct beam is retarded by 90°, the features appear brighter than the reference constituent, since now OA' and $A'B'$ are additive, and the effect is described as negative phase contrast.

Schematic diagrams of the arrangements for phase contrast microscopy in reflected light and transmitted light, which are identical in principle, are shown in Figs. 6.1 and 6.7*a* respectively. In the latter case, an annular diaphragm is located in the front focal plane of the condenser, so that it illuminates the specimen with a hollow cone of light. The condenser is focused so that an image of the annulus is formed in the back focal plane of the objective, i.e. the rays passing through the specimen are collimated. A transparent phase plate, that carries an annulus the same size as the image of the condenser annulus, is inserted in this plane and is made to coincide exactly with it. The annulus of the phase plate accelerates or retards the phase of the direct beam, all of which passes through it. Diffracted, refracted or reflected beams pass through the whole of the phase plate and most of them under go no phase change because they do not pass through the annulus. Thus the direct and diffracted beams interfere to form the image.

For maximum contrast in the final image the two beams must be of about equal intensity, thus the annulus of the phase plate is coated with an absorbent material, usually antimony, that reduces the intensity of the direct beam by about 80%, so that it matches that of the diffracted beam. The phase changing annulus may be either a groove in the glass that reduces the optical path length through the glass, thus advancing the direct beam relative to the diffracted beam, giving positive phase contrast (Fig. 6.7*b*), or a transparent layer of dielectric material, often magnesium

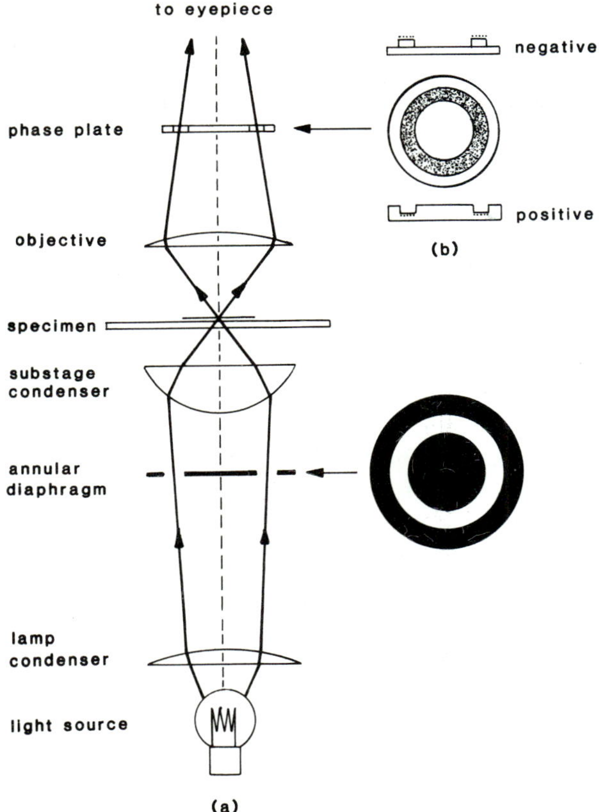

Figure 6.7 (*a*) Schematic diagram of the transmitted-light phase contrast microscope. (*b*) Sections through positive and negative phase plates; the shading represents the absorptive layer on the phase ring.

fluoride, to retard the direct beam, giving negative phase contrast (Fig. 6.7*b*).

Occasionally, the annuli of the phase plate and diaphragm have been replaced by a cross or slit. However, the annular arrangement is preferable because it gives symmetrical illumination.

With transparent specimens, positive phase contrast causes areas of greater optical density, i.e. greater R.I. or thickness, to appear darker than their surroundings and vice versa for negative phase contrast. Variations in optical thickness of about 3 nm can be detected using a positive phase plate in green light. The difference in optical thickness, t_0, is given by the equation

$$t_0 = (n_1 - n_2)t \qquad (5.1)$$

where n_1 and n_2 are the R.I.s of the two constituents and t is the physical thickness of the specimen. Thus if a constituent in a slice of material 30 μm thick is to be made visible by phase contrast, the difference between the R.I.s of the constituents needs to be only 10^{-4}. Thus the technique is useful for revealing subtle differences in R.I.s between constituents or regions within a specimens, as in organic polymers and glasses.

Unfortunately, the images formed are imperfect. The most pronounced blemish is the formation of bright haloes around constituents, owing to the direct and diffracted beams being imperfectly separated.

In reflected light, phase contrast microscopy enables differences of surface level of 5–50 nm to be distinguished, maximum contrast being obtained in the range 20–30 nm. However, absorption and the halo effect may confuse the image. True phase contrast effects are identified by examining the feature in turn under positive and negative phase contrast conditions. If the feature changes from light to dark, or vice versa, on changing from one to the other the effect is due to phase contrast and indicates a difference in level. Under positive phase contrast depressions appear darker than high spots and vice versa for negative phase contrast. If the surface levels of features differ by larger amounts, e.g. about 100 nm, the contrast may be reversed, but differences of this magnitude can be seen in bright-field illumination. In many cases the effects observed using phase contrast illumination depend more on the absorption of light by the phase ring than its ability to shift the phase of the light. Thus the image may be regarded as being formed by a combination of phase contrast and opaque stop illumination.

When we examine etched specimens using phase contrast illumination the etching must be very light if we are to obtain satisfactory results. Deeper etching may cause phase-contrast and normal bright-field images to be superimposed. Figure 6.3 compares the images of a lightly etched mild steel specimen examined using bright-field, opaque stop and positive and negative phase contrast illumination.

7 Interference microscopy

Interference microscopy makes use of the fact that coherent beams of light waves that are out of phase by half a wavelength interfere destructively. Indeed, phase contrast microscopy, discussed in the previous chapter, is a special case of it. The techniques of interference microscopy enable variations in surface level to be shown by interference fringe contours or by changes in contrast. Fringe patterns, called 'interferograms', are analogous to contour maps and enable much more accurate quantitative deductions to be made about surface topography than is possible with opaque stop or phase contrast techniques. The principles of the methods are simple and are discussed below.

7.1 Two-beam interferometry

Instruments that make use of the interference of two coherent beams of light may be divided into two groups: those that employ a pair of matched objectives and those that employ a single objective. The first group is typified by the Linnik interference microscope and the second by the Dyson interference attachment.

7.1.1 *The Linnik interference microscope*

One of the simplest arrangements for two-beam interferometry is that used in the Linnik interference microscope (Fig. 7.1). A monochromatic beam of light is divided by the beam splitter B into two beams of equal intensity, perpendicular to each other. One beam passes through a high-power objective O and is focused on the specimen S, while the other passes through a *matched* objective O' and is focused on a metallized optical flat F. The beams reflected from the specimen and optical flat meet at the beam splitter, having followed identical paths, except that F is optically flat and S is not. The beams from the two surfaces interfere destructively in the front focal

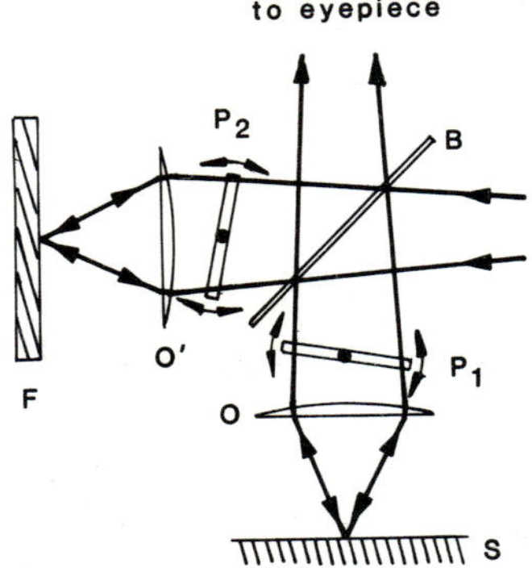

Figure 7.1 Schematic diagram of the Linnik interference microscope.

plane of the eyepiece if their phase difference is $n\lambda/2$, where n is an odd number. Fringes are produced that link together points at the same level, thus yielding a contour map of the surface if the reference flat is perfectly flat. The change in height of the surface is half a wavelength from one fringe to the next. The intensity of the light in the fringes varies as the square of the cosine of the phase difference (Fig. 7.2*a*). The half width of a fringe is the width of the intensity-distance curve at half maximum intensity, *PQ*, and the distance between successive fringes is twice this, *RS*. Thus, *if* the position of a fringe can be located to a *fifth of the half width*, the hatched area of width *d* in the figure, the maximum precision with which differences in level can be measured is $\lambda/20$, e.g. in green light of wavelength 500 nm differences of level of ~25 nm can be measured. However, with irregular surfaces this degree of precision is difficult to attain.

Special interference microscopes are used in practice, rather than attachments to conventional microscopes, because the arrangement must be mounted very rigidly and is not readily attached to a standard microscope. Also in practical arrangements rotatable glass plates P_1 and P_2 are included in the optical system. P_1 enables the path of the rays to the

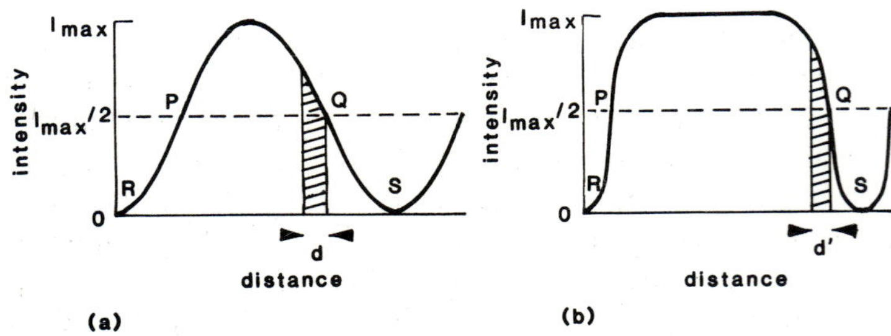

Figure 7.2 Intensity distribution in interference fringes. (*a*) Two-beam fringes; (*b*) multiple-beam fringes.

specimen to be adjusted, while P_2 is used to adjust the separation of the fringes.

7.1.2 *Single-objective two-beam interferometers*

A well-known single-objective interferometer is the Dyson two-beam interference attachment, which is mounted between the objective and the specimen. The attachment for the examination of opaque specimens consists of three glass blocks, 1, 2 and 3, cemented together and oiled to the specimen (Fig. 7.3). The incident monochromatic beam of light passes through the blocks, is reflected from the annular mirror A on the curved face of block 3, and is partially reflected at the interface between blocks 2 and 3. The partially reflected beam falls on the surface of the specimen and is reflected back into the attachment. Simultaneously, the beam partially transmitted at the interface between blocks 2 and 3, that acts as the reference beam, falls on the small central mirror M at the interface between blocks 1 and 2, which is the same distance from the partially reflecting interface (2, 3) as the specimen, and is reflected. The two beams recombine and form interference fringes at F, which are viewed by the objective.

A similar arrangement, due to Mireau, makes use of a partially reflecting plate midway between objective and specimen, and a reference mirror at the centre of the front lens of the objective. A third arrangement, due to Michelson, uses a beam splitter, similar to that in the Linnik interferometer, and a reference surface in front of the objective. The latter has the advantage over the other single objective interferometers that the reference surface can be tilted in order to alter the spacing of the fringes.

Similar arrangements are available for two-beam interference microscopy using transmitted light.

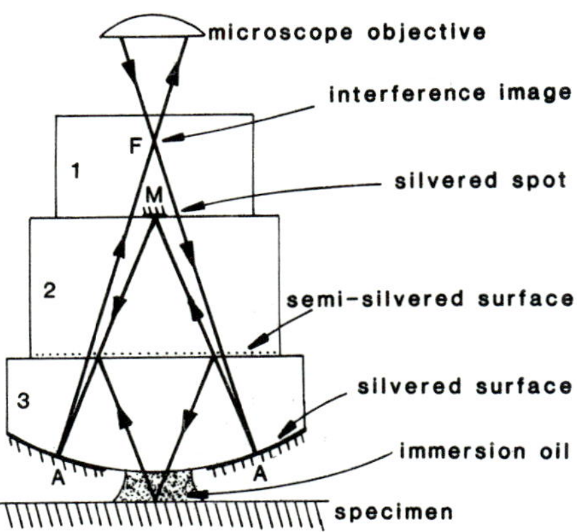

Figure 7.3 Schematic diagram of the Dyson interference attachment.

7.2 Multiple-beam interferometry

The technique uses simpler equipment than that needed for two-beam interferometry and is much more sensitive. A partially silvered reference flat is placed closely parallel to the specimen surface, and the specimen is illuminated by a well-collimated beam of monochromatic light at normal incidence (Fig. 7.4). The incident beam is reflected many times between the specimen and the reference flat. At each reflection from the specimen, part of the reflected beam emerges from the reference flat, producing a series of reflected beams that are parallel to the first reflected beam (Fig. 7.4). The reflected beams interfere and produce fringes in a similar manner to those produced in a two-beam interferometer. Dark fringes are formed when the distance between the reflecting surface at a specific point is either $\lambda/4$ or an odd multiple of it. Thus the thickness of the wedge between specimen and reference flat changes by $\lambda/2$ on moving from one fringe to the next.

However, the intensity distribution in the fringe pattern is more complex than that in a two-beam interferogram (Fig. 7.2*b*), and makes the fringes much sharper and narrower. This makes it possible to estimate the levels of features very accurately and to pick out small features. Under exceptionally good conditions measurement of differences in level of about a fifth of d^*, where d^* is 1/20 of $\lambda/2$ (Fig. 7.2*b*), is possible, i.e. displacements of about 1/100 of a fringe spacing are measureable. In green light of wavelength 500 nm this corresponds to a difference in level of 2.5 nm, i.e. about ten atomic diameters. However, under most practical conditions such accuracy is rarely attainable.

With a perfectly flat specimen a series of parallel, equally-spaced fringes would be seen. However, with a specimen with an undulating surface the fringes represent lines of equal wedge thickness, and thus are contour lines whose separation in level corresponds to $\lambda/2$. We can identify regions where the separation is a minimum by gently squeezing the specimen and flat together, when fringes move away from the points of minimum separation. The interferograms are interpreted in a similar manner to contour maps, e.g. very closely spaced fringes indicate a steeply sloping region.

In practice, to obtain maximum sensitivity, up to 100 reflections may be

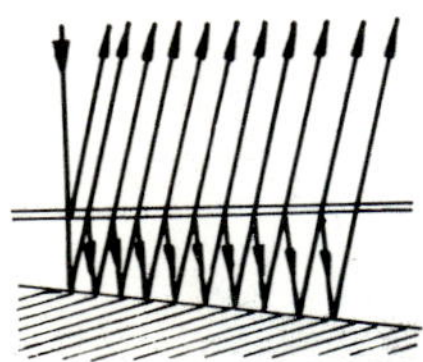

Figure 7.4 Multiple reflections in the multiple-beam interferometer.

needed, thus the specimen must be highly reflective and about 90% of the light must be reflected at the surface. Many well-polished metals will produce a dozen or more reflections, which produce sharper fringes than two beams. Nonetheless, to achieve the maximum sensitivity a uniform layer of silver that does not alter the surface contours is evaporated onto the surface in vacuum.

The reference plate must be optically flat, be sufficiently thin to go between specimen and objective, have a highly reflective upper surface, and must not appreciably absorb the light during multiple reflection. Selected* microscope coverslips make satisfactory reference plates when one surface is coated with an evaporated silver layer that has a reflectivity similar to that of the specimen. Silver films with a reflectivity of about 90% and a transmission of about 1% are suitable and are recognised by their purplish colour. Such films oxidize readily and may be protected by an evaporated silica layer. Alternatively, for some metallographic applications evaporated aluminium films work well and last indefinitely.

It is essential to use monochromatic light or, at worst, a very narrow range of wavelengths. The beam must be normal to the surface and highly collimated, because oblique incidence and divergence of the beam cause the multiply-reflected beams to spread to an unacceptable extent. Thus, with metallurgical microscopes, Köhler illumination *with the aperture stop closed down* is satisfactory, but the resolving power of the microscope is restricted.

7.3 Interference contrast

The techniques already described produce fringes that form a contour map of the surface examined. However, it is possible to widen the fringes so that only one or two occupy the field of view, e.g. by tilting the mirror in a two-beam interferometer. This alters their appearance and produces an effect called interference contrast that is similar to phase contrast. However, interference contrast can also be produced by the use of birefringent plates and polarized light. Attachments to polarizing microscopes are available and two, which are widely used, are described below.

All polarizing-interference microscopes make use of the duplication of the image in a birefringent material. Two slightly displaced images that differ in phase by $\lambda/2$ are produced. These annihilate each other at all points excepting those where the displaced images are out of step (Fig. 7.5) thus revealing the structure.

* Coverslips are selected by placing pairs face to face and observing them in monochromatic light. Provided that not more than a few fringes are seen they are considered to be sufficiently flat for reference plates.

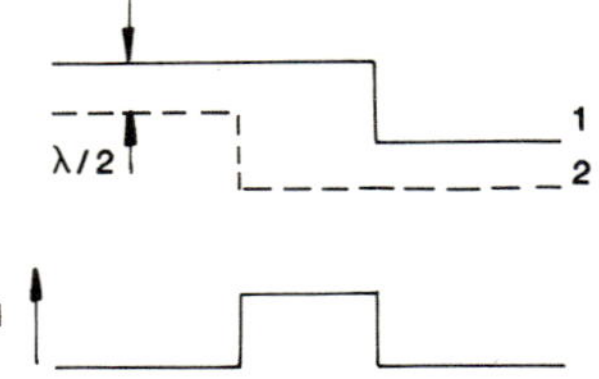

Figure 7.5 Schematic representation of the interference between slightly displaced images 1 and 2 that differ in phase by $\lambda/2$, producing variation in intensity I at the step.

7.3.1 *Françon's interference eyepiece*

In the eyepiece, image duplication is achieved by use of a Savart plate consisting of two quartz plates of identical thickness, both cut at 45° to the crystal axis and mutually crossed (Fig. 7.6*a*) and cemented together. The ordinary ray O_1 formed in plate 1 becomes the extraordinary ray E_2 in plate 2, while the extraordinary ray E_1 formed in plate 1 becomes the ordinary ray O_2 in plate 2. If the ray E_1 is displaced vertically with respect to ray O_1 in plate 1 (Fig. 7.6*b*), the ray E_2 (which was O_1 in plate 1) is displaced horizontally with respect to ray O_2 in plate 2 (Fig. 7.6*c*).

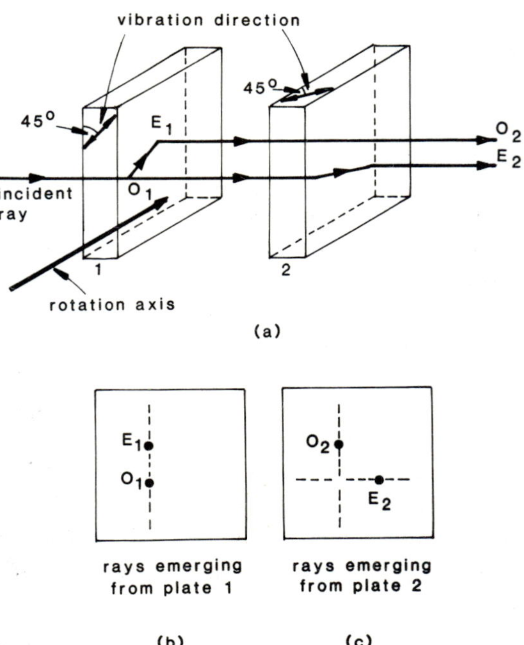

Figure 7.6 Principle of the Savart plate. The individual quartz plates that make up the Savart plate are shown separated for clarity.

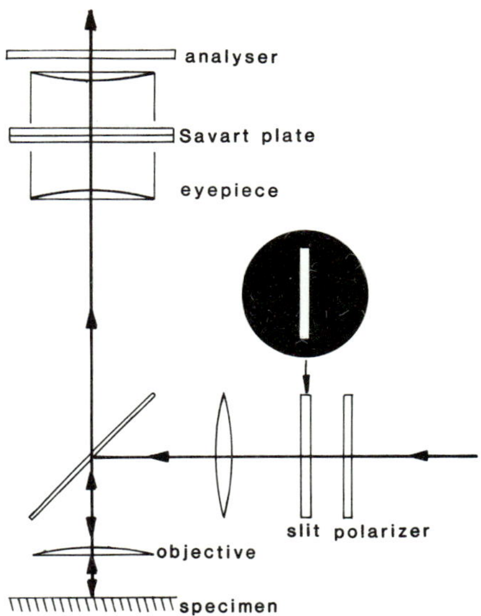

Figure 7.7 Schematic diagram of the Françon interference eyepiece arrangement for reflected light.

When the incident ray is normal to the Savart plate the emergent rays E_1O_2 and O_1E_2 are parallel and in phase, because their paths are of equal length and symmetrically arranged relative to the incident ray. However, when the plate is rotated about the axis shown (Fig. 7.6*a*) the paths of the two rays differ depending on the extent of the rotation. By using a beam from a slit and imaging it on the Savart plate, which is located in the eyepiece (Fig. 7.7), we can select particular fringes from the series of coloured fringes produced by white light. This controls the colour and contrast of the image.

7.3.2 *The Nomarski interference microscope*

In this case, the duplication of the image is produced by a parallel-sided double quartz prism, called a Wollaston prism, made by cementing together two quartz wedges of angle α, with their equivalent polarization directions perpendicular (Fig. 7.8*a*). It may be used with either transmitted or incident light polarizing microscopes, and the schematic arrangements of the microscope for both transmitted and incident light are shown in Fig. 7.8*b*.

Consider the incident-light case in which the light entering the eyepiece traverses the prism twice. For a ray polarized parallel to either direction of polarization of the prism, the optical thickness of the prism varies from edge

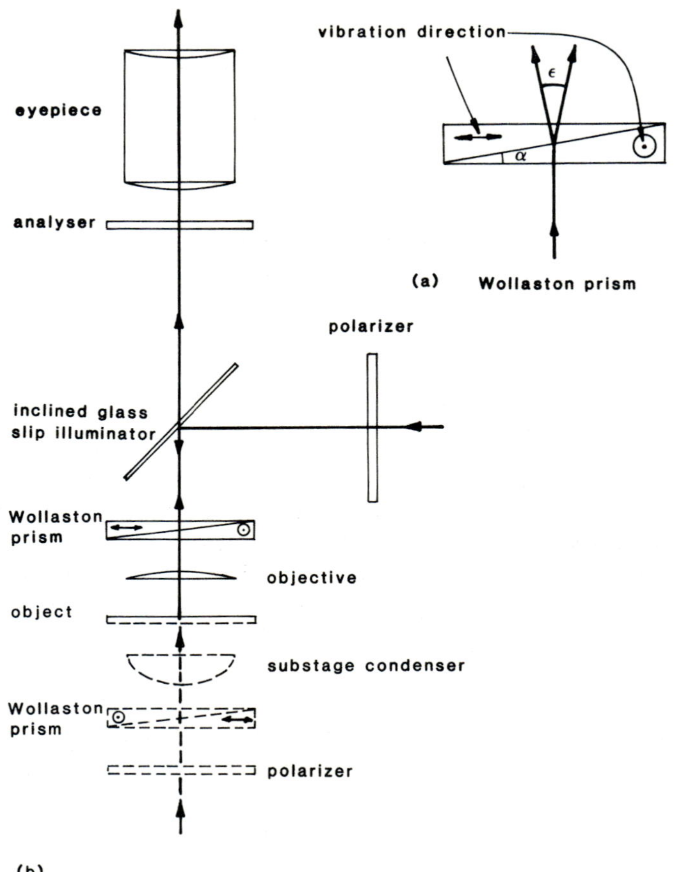

Figure 7.8 (*a*) Details of the Wollaston prism: the arrows show the vibration directions of the quartz wedges. (*b*) Schematic diagram of the Nomarski interference microscope. The arrangement for incident light illumination is shown by full lines; the alternative illuminating system for transmitted light illumination is shown by broken lines.

to edge continuously. But, because each ray traverses the prism twice, an incident ray passing through the prism a certain distance from the centre, after reflection passes through it at an equal distance from the centre on the opposite side. Thus because the total optical path is equal to twice the optical path through the centre all rays travel the same optical distance through the prism. This is necessary if the rays are to interfere coherently in the image plane. (A similar argument applies in the transmitted light case where there are separate Wollaston prisms in the illuminating system and microscope.)

The birefringent quartz divides the beam into two beams that are

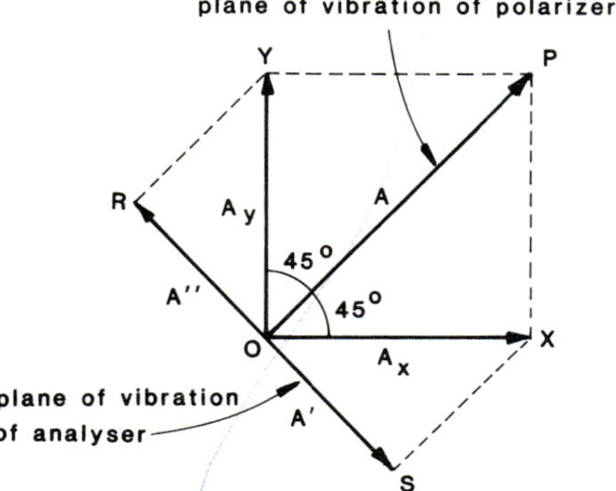

Figure 7.9 Vector diagram showing the effects of the Wollaston prism and analyser on the incident plane polarized beam.

polarized in planes perpendicular to each other and that are inclined at a mutual angle ε. Consequently two images are produced that are slightly displaced sidewise and mutually inclined at angle ε.

The polarizer, Wollaston prism and analyser act together to cause interference between the images, as shown by the vector diagram (Fig. 7.9). OX and OY represent the polarization directions in the quartz. The plane of polarization of the incident beam is parallel to OP and is at 45° to both OX and OY, consequently the amplitudes A_x and A_y of the two beams emerging from the prism are equal. These fall on the analyser, whose plane of polarization is at 90° to that of the polarizer, which passes components OR

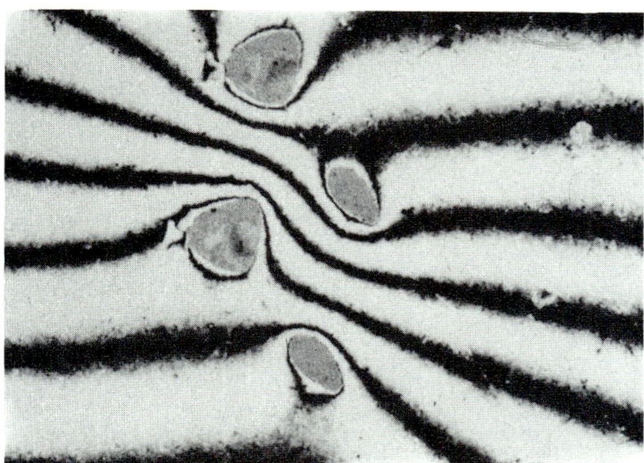

Figure 7.10 Photomicrograph showing the duplication of the image of a pair of small blowholes in a cast aluminium alloy (LM11) and the associated interference fringes produced by the Nomarski interferometer (× 125). When the specimen is rotated the pattern of images and fringes changes because the direction in which the images are displaced is fixed.

and *OS*, whose amplitudes A'' and A' are equal but opposite. Thus the rays forming the image are extinguished except where the detail is slightly displaced and, in general, a double image is formed, e.g. Fig. 7.10. The separation of the double image is determined by the angle α of the wedges in the Wollaston prism.

By rotating the analyser the conditions can be varied, thus certain wavelengths only may be extinguished yielding a coloured image. When the analyser is set at 45° relative to the polarizer its plane of polarization is parallel to either *OX* or *OY*, one beam is extinguished and a normal unduplicated image is formed. Removing the analyser does not eliminate the double image.

Because the detail in the double image is slightly displaced, interference fringes form. Their spacing is governed by the mutual angle of inclination ε of the two images and the position of the Wollaston prism. When the prism is moved along the optic axis of the microscope towards the back focal plane of the objective the spacings and widths of the fringes increase until the band between two adjacent fringes occupies the whole field. This permits the arrangement to be used either as an interferometer or to produce interference contrast.

7.3.3 *The Nomarski polarization interferometer*

Monochromatic light is used to permit the formation of sharp fringes, and their spacing may be adjusted by moving the Wollaston prism along the microscope axis. It is essential that the image lateral displacement be greater than the feature to be examined, so that interference can occur between the feature and its *plane* surroundings. When this condition is satisfied, differences in level can be calculated from deviations of the fringes. However, if the whole surface shows surface relief details in the image overlap and become confusing, and calculation of differences in level cease to be possible.

7.3.4 *The Nomarski interference contrast arrangement*

For interference contrast the angle α of the wedges in the Wollaston prism is chosen to give an image separation corresponding to about the limit of resolution, so that doubling of the image detail is not perceived. When this criterion is satisfied the point of intersection of the two beams lies outside the Wollaston prism. If the position of the Wollaston prism is adjusted so that the point of intersection lies in the back focal plane of the objective the band between adjacent fringes spreads uniformly over the entire image field. When in this position and using white light, the arrangement is very sensitive to slight differences in surface level that are revealed as differences in either intensity or colour, giving effects similar to phase contrast.

7.4 Applications

Interferometry has been used in the study of surface topography of opaque specimens. Such studies have yielded useful information about deformation processes including slip and twinning at room temperature, slip and grain boundary sliding in elevated temperature creep, and deformation and fracture in fatigue. Moreover, it enables accurate measurements of step heights and surface tilts in individual grains, and shows anisotropy of deformation. It is also used in the study and measurement of fine-scale surface roughness, down to ~ 3 nm, in the measurement of the thickness of thin surface layers, and to reveal details of cleavage facets and similar features in crystals. Moreover, the technique is also applicable to transmitted light microscopy and has found applications in the biological field.

Interference contrast microscopy is useful for enhancing contrast in specimens and is a superior alternative to phase contrast microscopy, because it does not give rise to the halo and absorption effects that mar the phase contrast images. Moreover, it offers a wider range of contrast condition, ranging from colour contrast, to which the eye is very sensitive, to very marked black and white contrast. The application of the Nomarski interference contrast technique is well illustrated by a recent study of fatigue in nickel base alloys, in which slip band development and crack formation and propagation were elegantly shown using colour contrast (reference 13).

8 Quantitative microscopy

Despite the fact that the microscope is most often used to make qualitative observations, it is capable of being used to make a wide variety of quantitative measurements. Such measurements are becoming increasingly important because we are often concerned with establishing quantitative relationships between the properties and structures of materials. We can divide the quantitative measurements into two groups. Firstly, there are those measurements that yield a simple direct measure of a particular property or feature, some of which have been mentioned earlier, e.g. optical properties of crystals, angles between crystal faces or grain boundaries, surface topography, and case thickness on surface-hardened steels. Secondly, there are measurements that we can use to deduce and quantify the three-dimensional structures of materials, which often have an important influence on their properties and behaviour, e.g. grain size, volume fraction of constituent, etc. Study of three-dimensional structures is called stereology and in this chapter we shall concern ourselves with some aspects of it.

The measurements, which are statistical in nature, can be made manually, but often the techniques are tedious because large numbers of measurements must be made. Fortunately, many of them are capable of automation and several instruments are available commercially that, *once they have been set up correctly*, will carry out speedily the numerous measurements required for an acceptable degree of accuracy. Such instruments are invaluable where measurements are to be made on large numbers of specimens. Nevertheless there are many occasions on which the manual techniques will yield the desired information without undue effort or where the instrumentation is incapable of distinguishing between constituents, e.g. in microstructures of sintered partially prealloyed steels.

8.1 Measurement of volume fraction of a constituent

It has been established mathematically that in a material in which the constituents are randomly distributed the volume fraction of a constituent,

V_v, is related to measurements on a random two-dimensional plane through the material by the following equations:

$$V_v = \frac{V_c}{V_t} = \frac{A_c}{A_t} = \frac{L_c}{L_t} = \frac{P_c}{P_t} \tag{8.1}$$

where V_c = volume of constituent, V_t = total volume of material, A_c = area of constituent, A_t = total area, L_c = length of line intercept through constituent, L_t = total length of line, P_c = number of points in constituent, and P_t = total number of points. Thus the volume fractions of constituents can be determined by areal analysis, lineal analysis, or point counting. Whichever method is used a satisfactory *statistical* coverage of the material must be achieved in order to obtain the desired degree of precision. Moreover, the planes examined must be representative of the material; if the structure is inhomogeneous or shows preferred orientation several *appropriate* sections must be taken. The data obtained is evaluated by statistical methods to obtain a mean value for the volume fraction and a measure of its precision, e.g. standard deviation.

8.1.1 *Areal analysis*

It follows from equation (8.1) that by measuring the areas of the grains of a constituent in a number of photomicrographs and the total area of the micrographs we can find the volume fraction of the constituent. We can make area measurements using a planimeter, or by cutting out the areas of the constituent in the micrograph and weighing all the pieces, or by superposing a square grid and counting the whole and fractional squares occupied by the constituent. The method is more time-consuming than the alternative methods, thus the latter are usually preferred.

8.1.2 *Lineal analysis*

This makes use of the second equivalence in equation (8.1). Random lines, say at angles of $\sim 15^\circ$ to each other, are superimposed on the microstructure, and the lengths of the intercepts through each of the constituents are measured. It follows that

$$L_t = L_\alpha + L_\beta + \cdots + L_\omega \tag{8.2}$$

where L_α, L_β, etc. are the intercept lengths of the constituents α, β, etc. Thus, no matter how many constituents are present, their volume fractions can all be determined. In practice, we carry out the analysis by drawing random lines on photomicrographs and measuring the intercept lengths, or by measuring intercept lengths with a scale in the eyepiece, or by traversing a specimen under the cross-wires of a microscope eyepiece using a micrometer stage and noting the lengths of the intercepts. A well-known mechanized form of the latter procedure is the Hurlbut counter. The method is not very satisfactory for finely-dispersed constituents because the lengths of very short intercepts are difficult to measure accurately.

8.1.3 *Point counting*

In this case we superimpose a grid on the image of the structure and count the number of intersections, i.e. points, falling within the desired constituent, P_c, and compare it with the total number of intersections, P_t. We count a point falling on a boundary as half. The method is particularly satisfactory for estimating the amounts of finely-dispersed constituents. We apply the method either by superimposing a grid on clear plastic sheet on photomicrographs or projected images, or by use of a reticule with a grid (Fig. 8.1) in the eyepiece. The grid preferably should have not more than about 25 intersections, because errors in counting are more likely when the number is large. For a given count, better precision is obtained using a regular grid than a random array of points. Moreover, the magnification should be selected so that on average only one point falls in each particle of constituent.

All these methods measure the *volume* fraction of a constituent. Weight

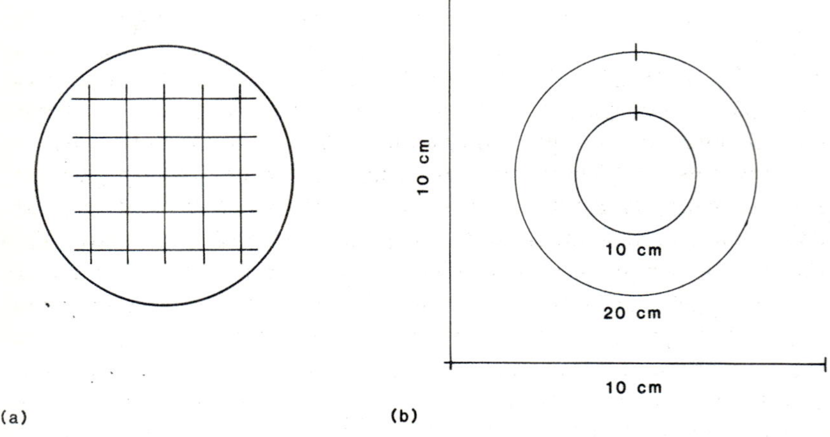

Figure 8.1 (*a*) Twenty-five-point grid eyepiece reticule for point counting. Alternatively, similar grids on transparent plastic can be superimposed on projected images of photomicrographs. (*b*) Hilliard's 10 cm and 20 cm circumference test circles for grain size measurement. The circles on transparent plastic sheet are superimposed on projected images or photomicrographs. The straight lines permit checking of dimensional changes during reproduction. Similar test circles can be produced as eyepiece reticules, but require calibration with a stage micrometer. (J.E. Hilliard, *Metal Progress*, 1964, **85**, 5, 99).

fractions, W_w, which are useful, for example, if we wish to accurately relate the fractions to phase diagrams, can be calculated from the volume fraction if the densities of the constituent, ρ_c, and the material, ρ_m, are known, thus

$$W_w = (V_v \times \rho_c)/\rho_m \tag{8.3}$$

8.2 Grain size and grain surface area

The simplest method of grain size measurement is to match the grain size of the material at a specified magnification with that of a similar standard material of known grain size on an ASTM grain size chart. The grain size is designated by the ASTM grain size number, N, of the standard that is closest in grain size to the specimen. N is defined by the equation

$$n = 2^{N-1} \tag{8.4}$$

where n = the number of grains per square *inch* of image when viewed at a magnification of × 100. On this scale the grain size changes by a factor of two when N changes by one; moreover, the larger is N the finer is the grain size. ASTM grain size numbers are widely used for metallic materials, especially steels and brasses, and give an acceptable, but not very precise, measure of grain size for many commercial purposes.

Of the various criteria that may be used to characterize grain size the most useful is the mean linear intercept, L. We determine the number of grain boundaries intercepted in a known length of traverse by superimposing on the image either lines of known length in random directions, or Hilliard's circular test figure, Fig. 8.1*b*, and counting the number of boundaries intercepted by the lines or circles. Hilliard's method is especially effective because the use of a circular figure takes into account the effects of orientation and there is no 'end of line' error, because a circle is endless. In practice, we select a magnification at which 6 to 10 boundaries are intercepted by the circle and for a precision of about ±0.3 grain size numbers we apply the circle until about 35 intersections are obtained. However, higher precision than this is necessary if we wish for example to correlate grain size to properties, and we must count a greater number of intercepts to attain it. The number of grain boundaries intercepted, P, is equal to the number of grains in Hilliard's method and approximately so when using random lines. Thus

$$L = \frac{\text{total length of lines}}{\text{total number of intercepts}} = \frac{l}{PM} = \frac{1}{N_1} \tag{8.5}$$

where N_1 = number of intersections per unit length, l = length of test line, and M = magnification. L is related to the *equivalent* ASTM grain size number G by the equation

$$G = -10.00 - 6.64 \log_{10} L \tag{8.6}$$

8.2.1 *Surface area of grains*

The surface area per unit volume, S_v, of grain boundaries and constituents is calculated from the mean linear intercept, L, used to characterize the grain size using the equation

$$S_v = 2/L = 2N_1 \tag{8.7}$$

The expression is valid for any space-filling arrangement of grains and is independent of the shape and disposition of the microstructural features provided that the plane of section is truly random.

The number of grains per unit volume, N_v, is related to S_v by the equation

$$S_v = KN_v^{1/3} \tag{8.8}$$

where K is a constant which depends on the grain shape; for equiaxed grains K is approximately 8/3.

8.2.2 *Inclusion counts*

Characterization of the 'dirtiness' of metallic materials, especially steels, by counting the number of inclusions in a prescribed area is widely used. Early methods employing comparison charts were very inaccurate. Recently, satisfactory automated image-analysing techniques have been developed that measure rapidly the number, size, size distribution, and types of inclusion.

8.2.3 *Particle size and spacing*

Often the relationships that relate to structures containing particles are valid for the general case of two constituent structures. An important relationship is the mean free distance of particles, λ, i.e. the average edge-to-edge distance of particles, that is given by the equation

$$\lambda = \frac{1 - (V_v)_p}{N_1} \tag{8.9}$$

where $(V_v)_p$ = volume fraction of particles, and N_1 = number of particle interceptions per unit length of test line. $(V_v)_p$ is determined, preferably by point counting, as described previously, while N_1 is obtained from counts of the number of particles intersected by random straight lines.

The mean random spacing, σ, i.e. the mean centre-to-centre spacing, is given by the equation

$$\sigma = \frac{1}{N_1} \tag{8.10}$$

Clearly, when the particle size becomes very small λ tends to σ. Furthermore, the mean intercept length, L_p, i.e. the mean particle diameter, is given by the equation

$$L_p = \sigma - \lambda = \frac{(V_v)_p}{N_1} \tag{8.11}$$

8.3 Accuracy

The measurements of the properties discussed in this chapter are statistical in nature. Thus, the accuracy of a measured quantity depends on the number of individual measurements made, but it is often difficult to predict how many measurements we need to make to achieve a given degree of accuracy. Nevertheless, in the cases of point counting and lineal analysis, methods of estimating the number of measurements to be made in order to attain a specified degree of accuracy have been deduced.

For point counting (20), the individual measurements for a randomly distributed constituent conform to the binomial distribution. Thus, the number of points in the constituent, n, that need to be counted for a desired value of the standard deviation, σ, is

$$n = V_v(1 - V_v)/\sigma^2 \tag{8.12}$$

where p is the fractional proportion of the constituent.

The equation presupposes that we know V_v, which we are trying to measure. However, we can estimate V_v from the first several applications of the point counting grid, or by visual inspection, and hence estimate n. Similarly, for lineal analysis (21),

$$n = 2V_v^2(1 - V_v)^2/\sigma^2 \tag{8.13}$$

where, in this case, n is the number of intercept lengths in the constituent that need to be measured.

Moreover, we see from these equations that to increase the degree of accuracy by an order of magnitude, the number of individual measurements that needs to be made increases a hundredfold.

9 Specimen preparation

We can see the true structures of materials only if we examine properly prepared specimens. Preparations of such specimens requires considerable skill that is developed only with much practice. Usually, it involves first cutting a small specimen from more massive material, because microscopes cannot deal with very large objects and preparation of large specimens is extremely difficult, and then preparing a surface or slice that can be examined under the microscope. The methods for preparing specimens of the different classes of materials are described below.

9.1 Metallic specimens

These may be prepared by one of the following methods: grinding and mechanical polishing, electropolishing, chemical polishing and, for very soft materials, microtomy. The latter method, which is rarely used, is described in connection with the preparation of polymer specimens.

9.1.1 *Sampling*

Representative specimens are cut from the object to be examined either by sawing or by using a thin abrasive cutoff wheel cooled with either water or water containing a water-soluble cutting oil. In all cases care must be taken to minimize heating and subsurface distortion, which may alter the structure of the specimen, by avoiding excessive pressure during cutting. Hard cutoff wheels should be used for cutting soft materials and vice versa. For cutting very hard materials, abrasive or diamond wheels are essential.

9.1.2 *Preliminary preparation*

Sawn surfaces are uneven and are flattened and smoothed by rubbing the specimen on a file held in a vice, not vice versa, by turning or by grinding on a fairly coarse grinding belt or grinding paper, e.g. 150 mesh. During grinding the specimen is kept cool by frequently dipping it in water. Slight rotation of the specimen during grinding helps produce a flat surface. Despite the fact that specimens cut with an abrasive cutoff wheel are flat, it is advantageous to grind them to remove any distortion that the wheel has produced. Flattening is preferably done before specimens are mounted in plastic, etc.

9.1.3 *Mounting*

Small and awkwardly shaped specimens are difficult to hold during grinding and polishing and usually they are hot mounted using a mounting press, at a temperature of $\sim 150°C$ and under pressure of 15–$30\,Nmm^{-2}$, either in a thermosetting plastic, e.g. phenolic resin, or a thermosoftening plastic, e.g. acrylic resin. Depending on the die used, mounting presses produce mounts of 25–50 mm dia. If the edges of the specimen need to be preserved, a plastic containing a filler of suitable hardness is used, or the specimen may be electroplated with a material of similar or greater hardness.

If hot mounting may alter the structure of the specimen, it is embedded in a cold-setting resin, e.g. epoxy, acrylic or polyester resin. In this case the specimens are placed flat face downwards on a greased flat glass plate surrounded by a metal ring and a mixture of liquid resin and catalyst is cast around them. The resin sets on standing for several hours at room temperature or more quickly at slightly elevated temperatures. However, note that the temperature of some resins may rise appreciably during setting.

Less frequently used alternative methods of mounting are embedding in low melting-point alloys, e.g. Wood's metal, or other low-melting-point materials, and holding specimens in mechanical clamps. The latter method is particularly suitable for polishing the edges of several pieces of sheet material simultaneously.

Porous materials, e.g. sintered products, must be impregnated with cold-setting resin before mounting or polishing, to prevent entrapment of grit, polishing media or etchant in the pores during preparation. This can be simply achieved by placing the specimens in a dish in a vacuum desiccator together with a small cup of liquid resin and catalyst, evacuating the desiccator and then immersing the specimens in the resin. When the vacuum is broken, the air forces the liquid resin into the pores. Then the specimens my be mounted in cold-setting resin as described before.

To prevent the specimen or mount rocking during subsequent fine grinding and polishing, the thickness of the specimen or mount is made about half its diameter. Moreover, the edges of specimens or mounts should be rounded, e.g. on a grinding belt or emery paper, because this minimizes the risk of damage to grinding papers and polishing cloths and of sore fingers.

9.1.4 *Fine grinding*

Smoothed specimens are ground successively on water- or paraffin-lubricated emery or silicon carbide grinding papers of progressively finer grit size; a widely used grinding sequence employs papers of grit sizes 180,240,400, and 600 mesh. On the first paper in the sequence the specimen is rubbed so that all the scratches run parallel to each other, and are roughly perpendicular to any scratches already on the specimen. This enables us to see clearly when all the previous scratches have been eliminated. When this is achieved the specimen and our hands are cleaned to remove particles of grit. Then the specimen is turned through about a right angle and ground on the next finest paper until the previous scratches are eliminated, and so on until the specimen has been ground on all the papers. After completing the grinding sequence both the specimen and our hands are thoroughly cleaned in water to remove residual grit, and then the specimen is rinsed in alcohol and dried in warm air; cleaning specimens in an ultrasonic cleaning bath may be helpful but is not essential.

9.1.5 *Lapping as an alternative to fine grinding*

Despite the facts that metallographic specimens can be ground by lapping and that lapping gives greater flatness, improved edge preservation, better retention of inclusions and more uniform surface level between constituents of different hardness, grinding of specimens using grinding papers has been preferred. This has occurred because lapping is wasteful of abrasive particles since they are not held by the surface of the lap, as are the particles in grinding papers. In particular, this has precluded the use of diamond powder for lapping, which is very efficient, because of its cost. Similarly, the use of diamond powders in grinding papers has been precluded by the inefficient use of the abrasive particles in papers.

However, there has been recently developed and is available commercially a plastic lapping wheel filled with metal powder that allows diamond particles to sit in the wheel surface without penetrating it and to be retained by it. The use of diamond on this kind of wheel permits rapid removal of material and the production of a plane surface simultaneously without waste of diamond powder, thus combining the advantages of both grinding and lapping. In practice, diamond powder is sprayed onto the wheel surface as a suspension in alcohol and an appropriate low viscosity oily lubricant is applied, and as it is used up new powder can be added. By using 6μm diamond spray, specimens that have been rough-ground to produce a plane surface can be fine-ground ready for polishing in a single operation. It is claimed that by using this technique there is considerable saving in cost compared with grinding on silicon carbide or emery papers, and preparation times are appreciably shortened.

9.1.6 *Polishing*

The finely ground specimens are polished on cloth-covered laps impregnated with fine abrasive powder and lubricant. Usually the laps either rotate or vibrate; rotating laps often rotate at a few hundred rpm, although faster speeds are preferable for certain materials. However, if desired, polishing can be done by hand on a stationary lap. During hand polishing on a rotating lap the specimen is rotated counter to the rotation of the lap under a light pressure that is learned by experience, and polishing is continued until all previous scratches are eliminated. On automatic polishing machines, which are used when many specimens need to be polished, and which use either rotating or vibrating laps, pressure is applied to the specimens by means of weights.

After eliminating all the previous scratches the specimen and our hands are thoroughly cleaned as described previously, and then the specimen is polished on another lap impregnated with a finer abrasive, and so on until the desired standard of polish is attained. Typically polishing is done on up to three grades of abrasive, e.g. 6, 1 and $\frac{1}{4}$ μm diamond particles. Selection of the most appropriate polishing conditions for a given material, i.e. abrasive, cloth, lubricant, lap speed and pressure, is a matter of experience.

Sometimes polishing is facilitated by use of a polish-etch attack, in which the specimen is polished using a slurry of abrasive with dilute etching solution. The strength of the etching solution must be sufficiently low so that the polishing action of the abrasive removes the effects of the etchant as they are produced, while the etchant removes most of the disturbed surface layer produced by the abrasive. For example, this procedure is used in the polishing of copper, silver and their alloys by using a dilute ammonia solution as lubricant; commercial brass and silver polishes often contain ammonia and are good for polishing specimens of these metals. However, the technique is not used with diamond abrasives, because they employ oily lubricants.

Electromechanical polishing is similar in concept to polish-etch attack. The specimen is polished on a conducting polishing wheel covered with a lap saturated with or immersed in an electrolyte, and the wheel is made the cathode and the specimen the anode in an elctrolytic cell. Consequently, polishing and electrolytic etching occur simultaneously. The technique is excellent for polishing copper base alloys and a variety of hard and precious metals.

9.1.7 *Polishing media*

The commonly used polishing media are alumina (aluminium oxide), magnesia (magnesium oxide), chromic oxide, ferric oxide and diamond. The oxide abrasives are used with an aqueous lubricant, but diamond must be used with the special oily lubricants recommended by the manufacturers. In general, the abrasive is worked into the lap with lubricant before polishing. During polishing, oxide abrasives (but not diamond) require replenishing frequently.

Diamond pastes (with either water-soluble or oil-soluble carriers) and sprays are available with particle sizes to $\frac{1}{4}\,\mu$m. Use of diamond is popular because specimens containing both hard and soft constituents can be polished without developing appreciable relief if an appropriate cloth is used, and because it is economical since laps need dressing infrequently. Moreover, use of diamond is essential for polishing very hard materials.

Alumina is widely used for general metallographic polishing. α-Alumina powders are available in particle sizes in the range $15 - 0.3\ \mu$m for initial polishing and γ – alumina powder with a particle size of $\sim 0.05\ \mu$m for final polishing.

Magnesia is used particularly for the fine polishing of aluminium and magnesim alloys. However, only alkali-free metallographic grades should be used, because soluble alkalis may stain the specimens. Moreover, the powders react with water and atmospheric carbon dioxide to form magnesium carbonate, which is likely to produce coarse scratches on the specimens. Thus cloths that are to be used with magnesia must be stored in a solution of 2% hydrochloric acid in distilled water; contaminated cloths can be restored by use of the same solution.

Chromic oxide and ferric oxide are available in a range of particle sizes similar to those for alumina. They are used mainly for polishing irons and steels.

9.1.8 *Polishing cloths*

The function of the polishing cloth is to hold the abrasive and apply it to the surface to be polished. It should have a long life and be free from both particulate and chemical contaminants arising from the manufacture of the cloth that may scratch or react with the specimen. Cloths are either stretched tightly over the polishing wheel and clamped or stuck to the wheel with a contact adhesive.

For rough polishing, cloths with little or no nap are preferred because they give better specimen-abrasive contact that produces fast cutting with minimal surface relief. The cloths most widely used are canvas, low-nap cotton, silk, nylon and nylon-rayon mixtures. In use the cloths are dressed with abrasive of the desired particle size and the wheels are rotated at speeds of 500–600 rpm for oxide abrasives but ~200 rpm for diamond.

Medium- or high-nap cloths are used for intermediate and final polishing. Commonly used cloths are felt, billiard cloth, selvyt, and special cloths with tightly packed upright synthetic fibres, although for polishing some metals special types of cloth are preferred, e.g. satin. Usually for the later stages of polishing wheel speeds up to ~300 rpm are used.

9.1.9 *Electropolishing*

However carefully done, mechanical polishing produces a layer of mechanically disturbed material on the surface that may be unacceptable in some applications. In such cases an undisturbed surface may be produced by either electropolishing or chemical polishing.

In electropolishing a masked or mounted specimen that has been ground to say 400 grit paper is made the anode in an electrolytic cell (Fig. 9.1). Using a suitable electrolyte and appropriate conditions of current, voltage and temperature, brightening and levelling occur during anodic dissolution, yielding a highly polished surface that is free from mechanical disturbance. Nevertheless, single-phase materials often polish more satisfactorily than do multiphase materials.

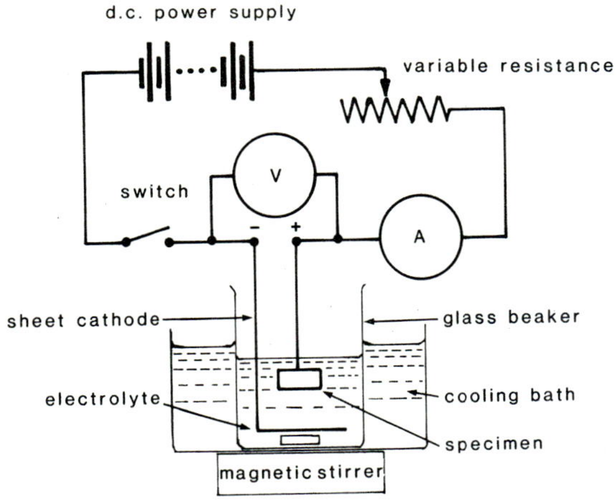

Figure 9.1 Schematic diagram of the arrangement for electropolishing and electrolytic etching.

On a microscopic scale we may regard a ground surface as a series of hills and valleys, superimposed on which are smaller irregularities. Both must be removed to produce a perfectly flat surface. An ideal polishing process therefore has two functions; it exerts a smoothing action by eliminating large-scale irregularities, say $> 1\ \mu m$, and a brightening action by removing smaller irregularities, say down to $0.01\ \mu m$. These functions are combined in electropolishing. The smoothing action occurs when an often-visible, relatively thick, viscous layer of reaction products forms around the anode during polishing. At high points on the surface the anolyte layer is thinner than in the depressions and permits more rapid dissolution of the former, thus smoothing or levelling the surface. Simultaneously, a thin film that controls the brightening action forms at the anode surface.

Typical anode current-cell voltage curves for low- and high-resistance electrolytes suitable for electropolishing are shown in Fig. 9.2. Over region *AB*, etching occurs, while from *B* to *C* both current and voltage fluctuate, then in region *CD* the anolyte layer forms and polishing occurs. Beyond *D* polishing accompanied by gas evolution and pitting or etching occurs. The optimum conditions for polishing occur at the point *P* on the curve where the tangent to the curve passes through the origin.

Low-resistance electrolytes are typified by the aqueous phosphoric acid or phosphoric-sulphuric acid solutions used for polishing copper alloys and stainless steels respectively, while high resistance electrolytes are typified by the perchloric acid-acetic anhydride mixtures used for polishing steels and many other metals. The compositions of numerous electrolytes and their operating conditions are reported in the literature; details are given in references 4, 8, 9, 23 and 24. Nevertheless, usually it is necessary to establish the best operating conditions experimentally when using an electropolishing solution.

Electrolytes containing perchloric acid are potentially very dangerous because they may react explosively with organic matter, e.g. plastic mounts.

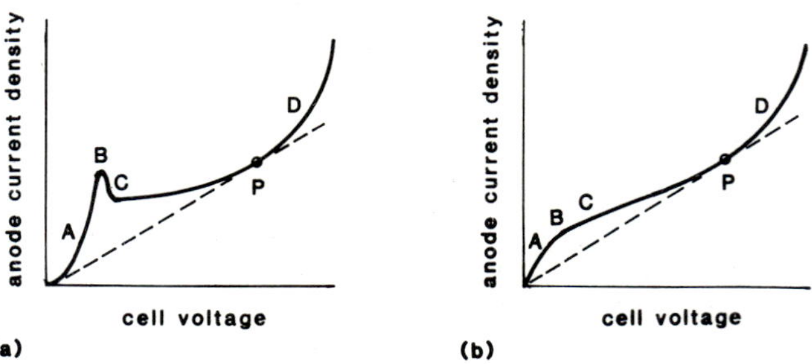

Figure 9.2 Anode current–cell voltage curves for (*a*) low-resistance electrolytes and (*b*) high-resistance electrolytes.

They should therefore be treated with great care both while preparing and using them, and the preparation and operating instructions should be followed as closely as possible.

9.1.10 *Chemical polishing*

Chemical polishing achieves the same effect as electropolishing but by purely chemical means. The polishing solution dissolves the surface layer of material, preferentially dissolving the high regions on the surface so that it is smoothed and brightened. The mechanism of polishing seems to be similar to that of electropolishing, but the technique is most suitable for use on single-phase materials. Many solutions have been formulated for the polishing of various metals; details of the compositions of some solutions are given in references 4, 23, and 24. For most metals the method is less satisfactory than electropolishing.

9.1.11 *Revelation of the structure*

After polishing specimens should first be examined in the polished condition. This reveals most clearly the presence of non-metallic inclusions, porosity, flaws, etc.; they are less easy to see and identify in an etched matrix. Also in some alloys, especially aluminium alloys, intermetallic compounds are clearly revealed. However, to see the structures of the metallic matrices specimens usually must be etched.

Etching embraces all techniques used to reveal the structures of the matrices. In pure metals and solid solution alloys etchants reveal the structure by either preferentially attacking grain boundaries or preferentially staining grains of different orientations, whereas in alloys with more than one phase they reveal the structure by preferentially attacking or staining one or more of the phases, owing mainly to differences in composition but sometimes also to differences in orientation. In addition, etching may be used to identify phases (references 4, 8, 9, 25–27) and inclusions (references 4, 28) and to indicate grain orientation and the positions of dislocations by etch-pit formation.

9.1.12 *Chemical etching*

Chemical etchants are often dilute solutions of acids or alkalis in water or alcohol, e.g. nitric acid in alcohol solutions, ‘nital’, used for etching steels, or such solutions to which oxidizing or other agents are added, e.g. ammonia and hydrogen peroxide solution used for etching copper-base alloys. A selection of commonly used etching reagents is listed in Table 9.1; very comprehensive lists of etching reagents are given in references 4, 8, 9, 25–27.

A specimen is etched either by immersing it in the etching solution or by

Table 9.1 Etchants for metals and alloys

No.	*Material*	*Etchant composition*	*Remarks*
1	Aluminium-base alloys.	$\frac{1}{2}$–2 ml HF, 100 ml H_2O*	Immerse or swab, 15–45 s.
2	" " "	1 ml HF, 1.5 ml HCl, 2.5 ml HNO_3, 95 ml H_2O*	Keller's reagent. Must be freshly made. Immerse or swab, 8–10 s.
3	" " "	1g NaOH, 100 ml H_2O	Swab, 5–10 s.
4	Copper-base alloys.	2–5 g $FeCl_3$, 5–30 ml HCl, 100 ml H_2O	Immerse or swab, 5–15 s. Grain contrast etch.
5	" " "	5 g $FeCl_3$, 5–30 ml HCl, 100 ml C_2H_5OH	Carapella's reagent. Immerse or swab, 10 s–several min. Grain contrast etch.
6	" " "	50 ml NH_4OH, 20–50 ml H_2O_2 (3%), 50 ml H_2O	Immerse or swab, 60s. Must be freshly made.
7	Aluminium and beryllium bronzes.	1 g CrO_3, 100 ml H_2O	Electrolytic etch, 6 V, 3–6 s, Aluminium cathode.
8	Iron, plain carbon and low alloy steels, cast irons.	1–5 ml HNO_3, 100 ml C_2H_5OH or CH_3OH (commonly 2 ml HNO_3)	Nital. Immerse, 5–30 s.
9	" " "	5 g picric acid, 100 ml C_2H_5OH	Picral. Immerse, 5–30 s.
5	Stainless steels.	As etchant 5.	Immerse or swab, 5–120 s.
10	" "	10 g oxalic acid, 100 ml H_2O	Electrolytic etch, 6 V, 10–15 s.
11	" "	10 ml HCl, 90 ml C_2H_5OH	Electrolytic etch, 6 V, 10–30 s.
12	Lead-base alloys.	10 ml acetic acid, 10 ml HNO_3, 40 ml glycerol	
8	" " "	As etchant 8: 5% nital.	
8	Magnesium-base alloys.	As etchant 8.	Immerse or swab.
13	" " "	2 g oxalic acid, 100 ml H_2O	Immerse.
14	" " "	2–5 ml acetic acid, 100 ml H_2O	Immerse.
15	Nickel-base alloys.	50 ml HNO_3, 50 ml acetic acid	Must be freshly made. Immerse or swab.
16	" " "	35 ml HCl, 5 ml H_2O_2, 60 ml H_2O	Swab.
17	" " "	5 ml H_2SO_4, 95 ml H_2O	Electrolytic etch. 1.5–4.5 V, 5–15 s.
18	" " "	5 ml acetic acid, 10 ml HNO_3, 85 ml H_2O	Electrolytic etch. 1.5 V, 20–60 s.
19	Tin-base alloys.	1 ml HCl, 100 ml C_2H_5OH.	Immerse.
1	Titanium-base and Zirconium-base alloys.	As etchant 1.*	Immerse. 5–10 s. Grain contrast etch.
20	" " "	1–2 ml HF, 100 ml saturated oxalic acid solution, trace $Fe(NO_3)_3$*	Immerse. 5–10 s. Grain contrast etch.
21	" " "	1–2 ml HF, 8–12 ml HNO_3 90 ml H_2O*	Immerse. 5–10 s. Bright etch.
8	Zinc-base alloys.	As etchant 8; 1 ml HNO_3.	Immerse.
19	" " "	As etchant 19.	Immerse.

* Hydrofluoric acid can cause serious burns that are difficult to heal. Rubber or plastic gloves must be worn when handling it, even in dilute solutions. If it is splashed on the skin or eyes it must be washed away immediately with water and *medical assistance obtained.*

Unless stated otherwise, concentrated acids must be used in making up etchants.

swabbing it with a small cotton-wool pad soaked with the etching solution until the structure is revealed. During etching the surface of the polished specimen becomes slightly dull, and with experience the required depth of etching is judged from the appearance of the surface. When the specimen is sufficiently etched it is washed in water or alcohol to stop the action of the etchant, rinsed in alcohol and dried in a stream of warm air. It is then ready for examination. If the structure is insufficiently revealed it is etched for a further time and re-examined. Nevertheless, if there is doubt about the depth of etching it is etched lightly and, if necessary, the etching is repeated until the desired depth is attained. However, if a specimen is overetched, it must be either repolished and re-etched or reground, repolished and re-etched, depending on the extent of overetching.

In many cases after initial preparation and etching the specimen surface is not completely free from distortion and scratches. It can be improved by alternately repolishing and re-etching until the desired quality of preparation is achieved. However, during the process the specimen should never be etched deeply.

Sometimes a particular etchant may not produce satisfactory etch on a specimen. In these cases the specimens are repolished to remove the etch, and an alternative etchant is used.

9.1.13 *Electrolytic etching*

In electrolytic etching the specimen is made the anode in an electrolytic cell and a relatively insoluble material is used as cathode, e.g. platinum or stainless steel. The experimental arrangement and procedure are similar to those for electropolishing, but the voltages and current densities are lower and etching is usually done at room temperature. Usually direct current is used but with some precious metals and their alloys alternating current must be used.

If we wish to electrolytically etch mounted specimens, the mount should be non-conducting and inert towards the electrolyte; if the mount is conducting it must be masked by use either of a lacquer or good-quality plastic insulating tape. Similarly, the faces of unmounted specimens that we do not wish to etch should be masked.

Electropolishing solutions can be used for etching but many solutions for eletrolytic etching are not polishing solutions. Some examples of electrolytic etchants are included in Table 9.1, and more comprehensive information is given in references 4, 8, 9, 25–27.

9.1.14 *Heat tinting*

This little-used method makes use of the fact that when heated in air different constituents oxidize at different rates owing to differences in composition and orientation. Thus at a given temperature thin layers of oxide of different thicknesses form on the various constituents, enabling them to be distinguished by differences in their interference colours.

In practice, well-polished specimens that have been thoroughly cleaned to remove surface contaminants and sometimes etched to remove the flawed surface layer are heated to a moderate temperature, rarely exceeding 400°C, on a hot plate or a sand bath, or in a bath of molten tin. As with conventional etching, the development of the tint is judged by observation. When the required contrast has been achieved the specimen is rapidly cooled in mercury, taking care not to allow the mercury to contact the oxidized polished surface.

9.2 Ceramic and mineral specimens

Three types of specimen are used for the examination of these materials. They are thin sections, polished sections and polished thin sections, and the methods of their preparation are described below.

9.2.1 *Thin sections*

First a thin slice ~ 5 mm thick is cut from the material using a diamond saw or cutting wheel. Porous or pliable materials may need to be vacuum-impregnated with a resin, e.g. epoxy resin, before cutting. Then one surface of the slice is ground on a cast-iron rotating lap using liquid suspensions of silicon carbide powders, starting with coarse powder, e.g. 80–120 mesh, and ultimately finishing with fine 600 mesh powder. Between successive stages in the grinding process the specimen and hands must be thoroughly cleaned. After final washing and drying the slice's ground surface is cemented to a microscope slide with either Canada balsam or a synthetic resin, which have melting temperatures in the range 90–140°C, by melting the resin on the slide and pressing the slice onto the resin to squeeze out air bubbles and excess resin. After this the exposed face of the specimen is ground with coarse silicon carbide until the thickness is reduced to $\sim \frac{1}{2}$ mm, and then successively on finer grades until the desired thickness is approached, again cleaning the specimen between successive stages. Finally, the last 20–30 μm is removed by hand grinding. The standard thickness of thin sections for mineralogical specimens is 30 μm, but for ceramic specimens the thickness may be as little as 10 μm, owing to their finer grain size. When the desired thickness is achieved the specimen is washed and dried, and a cover slip is cemented onto its upper surface with resin.

The thickness of a specimen is judged by examining the specimen under the microscope between crossed polars from time to time during the final grinding stage, and observing the interference colours of certain minerals present in the specimens. The interference colours shown by a mineral depend both on its optical properties and its thickness, and they can be calculated or deduced from birefringence charts, e.g. silica shows the first-order yellow colour when 30 μm thick.

9.2.2 *Polished sections*

Polished sections are prepared by mounting, grinding and polishing as described for metallographic specimens. Alternatively, the *grinding* may be carried out as for thin sections. As with thin sections and porous metals, porous specimens need to be impregnated with resin before grinding. After polishing, specimens must be etched, and examples of etchants are given in Table 9.2; more comprehensive information about etchants is contained in references 4, 14 and 26. In general, ceramic specimens reflect poorly and often lack contrast, but the reflectivity and contrast of etched specimens can be enhanced by vacuum coating the surface with silver to a depth of 20–30 nm.

9.2.3 *Polished thin sections*

These differ from ordinary thin sections mainly in that the upper surface of the specimen is not covered with a cover glass, but is polished. Their preparation is as described for thin and polished sections, but particular care must be taken in polishing to prevent the section breaking up or curling, because the cements have low melting points, and they soften

Table 9.2 Etchants for ceramics

Material	*Etchant composition*	*Remarks*
Aluminium oxide	2 ml HF, 98 ml H_2O*	
Magnesium oxide	Phosphoric acid.	4 min.
" "	50 ml HNO_3, 50 ml H_2O	1–5 min.
" "	1 ml H_2SO_4, 99 ml C_2H_5OH	
Silicon dioxide	Phosphoric acid.	
" "	Concentrated HF*	1–5 s.
Basic refractories	10 ml HF, 90 ml H_2O*	5–60 s.
" "	Concentrated HF*	2–3 s.
" "	Concentrated HCl	2–3 s.
" "	0.1 ml HNO_3, 100 ml C_2H_5OH	5–15 s.
Slags	1 ml HNO_3, 99 ml C_2H_5OH	30 s.
Cement clinker	Water followed by 1 ml HNO_3, 99 ml C_2H_5OH	1–5 s each.

* Hydrofluoric acid can cause serious burns; see the note to Table 9.1.

readily if overheating occurs. They may be examined by both transmitted and reflected light microscopy, which is particularly useful if some constituents are opaque.

9.3 Polymer specimens

Both thin sections and polished sections of polymers may be prepared. The methods of their preparation are described below.

9.3.1 *Thin sections*

Often thin sections of organic polymers are prepared from solid material by cutting thin slices with a microtome. They need to be cut at temperatures below the glass transition temperature of the material, which may be more than 100°C below room temperature. A cut section curls up during cutting and must be unrolled and mounted on a microscope slide for examination. It is unrolled in a bath of liquid, e.g. xylene, at room temperature and mounted by sandwiching it between a microscope slide and cover slip that have been coated with a few drops of mountant.

The mountant must wet the specimen, be compatible with it, and the temperature at which mounting is done must not change its structure. Materials used for mounting include chlorinated hydrocarbons, glycol bori-borate, and aqueous polyvinyl alcohol solutions. The mountants used for ceramic materials are unsuitable for use with polymers.

The thicknesses of the cut slices of polymer lie in the range 2 to 30 μm, depending on the type of material and the examination technique to be used. Specimens containing large amounts of filler, or those to be examined by dark-field, interference contrast or phase contrast techniques should be thin. Thick specimens are more suitable for birefringence studies. Moreover, specimens may be stained to reveal detail in bright-field illumination.

An alternative method for thin-section preparation for harder materials,

Table 9.3 Etchants and solvents for plastics

Material	*Etch or solvent composition*	*Remarks*
Polyethylene	Xylol	70°C, 60 s.
Polypropylene	100 ml HNO_3, 100 ml dichromate-sulphuric acid	70°C, 120 s.
Acrylonitrile-butadiene-styrene (ABS)	40 ml H_2SO_4, 10 ml H_3PO_4, 10 ml H_2O, 2g CrO_3	70°C, 180 s.
Polyamide	Xylol	70°C, 60 s.
Polyoxymethylene	30 ml HCl, 70 ml H_2O	20°C, 20 s.
Polycarbonate	60–80 ml chloroform, 40–20 ml acetone	

especially filled materials and composites, is to cement a flat, grease-free specimen onto a glass slide with a cyanoacrylate adhesive that hardens in about 20s, and grind and polish it is the same manner as a polished thin section of a ceramic material until the desired thickness is achieved. Porous specimens must first be vacuum-impregnated with epoxy resin.

9.3.2 *Polished sections*

These are prepared in the same way as metallographic specimens, grinding first on a series of silicon carbide papers, e.g. 320, 800, and 1200 mesh papers, and then polishing with diamond powder and a suitable lubricant on a hard cloth. Elastomers are more difficult to polish than thermosetting polymers and require longer polishing times.

The kind of polishing cloth used is important, especially with elastomers, and best results are obtained with smooth, synthetic fibre cloths. The choice of lubricant is also important; lubricants that may be absorbed by the specimen causing it to swell must be avoided.

Crystalline regions in polymers can be shown by either etching in acids or dissolving the surface in a solvent. In both cases crystalline regions are attacked more slowly than amorphous ones, thus revealing the structure. Organic solvents dissolve polymers more readily than inorganic ones. Nevertheless, it is important to choose correctly the solvent for each polymer. Also the choice of working temperature is important because melting of the surface and too severe attack must be avoided. A list of etching reagents and solvents is given in Table 9.3; further information may be found in reference 4.

9.4 Special techniques

9.4.1 *Cathodic vacuum or ion beam etching*

When a specimen is made the cathode in a vacuum glow discharge, it is bombarded with positive ions that are sufficiently energetic to remove atoms from its surface. The rate at which the atoms are removed depends on the microstructure of the material, consequently after treatment the structure is revealed in the same manner as in conventional etching. In practice specimens need to be well cooled as considerable heat is generated in the discharge.

The technique has the advantages that it can etch different constituents simultaneously and is able to show very fine detail in both lightly and deeply etched specimens; fine detail is lost when specimens are deeply etched by conventional etchants. It is applicable to both metallic and ceramic specimens, and is particularly useful when there are no satisfactory conventional etchants, or where conventional etchants may modify some feature of the specimen, e.g. enlarge pores in porous specimens. Staining arising from seepage of etchant or washing fluid from pores is also eliminated.

9.4.2 *Interference layer microscopy*

This is a fairly recently developed technique in which a very thin layer, 30–40 nm thick, of a transparent material is evaporated or sputtered under vacuum onto the surface of a well-polished specimen. Light rays falling on the layer are partially reflected at the air-layer interface and at the substrate-layer interface, so that multiple reflections occur within the layer (Fig. 9.3). Differences in reflectivity and colour between the phases are enhanced by the layer. When for a given wavelength the waves are out of phase by $\lambda/2$, destructive interference occurs and the wavelength is extinguished. Thus the contrast of a constituent can be adjusted by selecting the wavelength of the light with which it is illuminated. Alternatively, when the specimen is illuminated with white light, colour contrast is obtained, the colour observed being complementary to that which has been extinguished.

Materials suitable for evaporated coatings have R.I.s in the range 1.35 to 3.25 and include cryolite, fluorite, cerium fluoride, lead fluoride, zinc sulphide, selenide and telluride, and titanium dioxide. Highly reflecting or strongly absorbing substrates need to be coated with films of very high R.I., > 2.5, but films of lower R.I. are satisfactory for subtrates that have low reflectivity or absorb weakly.

Sputtered layers are produced using metal cathodes and sputtering in an oxygen-containing atmosphere; thus the sputtered layers are believed to be oxides. The most widely used cathode material is iron, but gold, platinum, copper, nickel, etc. have also been used.

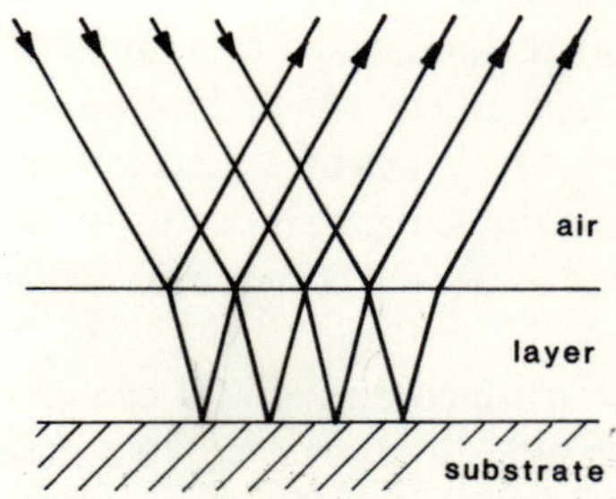

Figure 9.3 The principle of interference layer microscopy.

The technique can be applied to metals, ceramics and polymers; it is particularly useful for materials that resist chemical attack. Extensive details of the technique and its applications are given in reference 29.

9.4.3 *Replicas*

Occasionally it is necessary to examine specimens that are too large to fit on the microscope stage and from which smaller specimens may not be taken, or to examine periodically features that appear and develop on a surface, e.g. slip lines and cracks on tensile, creep and fatigue specimens. In these cases replicas are made of the specimen surface, using the techniques developed for electron microscopy, and are examined under the microscope.

A simple method for making replicas is as follows. Soak a suitably sized piece of clean cellulose triacetate film about 0.5 mm thick in acetone for several seconds to soften it. Using tweezers place it on the surface to be replicated and press it firmly into contact with the thumb or a squeegee. Allow it to harden for a few minutes and then gently peel it off starting at a corner. After the film is stuck with the replicated surface uppermost on a microscope slide with adhesive tape, it can be examined.

Often the contrast of the replica in incident bright-field illumination is adequate, but if desired it can be improved by vacuum deposition of aluminium, chromium or silver on its surface. Alternatively, phase contrast or interference contrast illumination can be used to enhance the contrast of an unmetallized replica. Cleanliness in the preparation of the replica is important, because the trapping of dirt or air bubbles beneath the film yields poor replicas.

One supplier of metallographic materials offers a green replicating tape, backed by a reflective coating, and a red solvent liquid, used in a manner similar to that described above. The replica is mounted with its reflective coating against the slide. The replicas are claimed to have reflectivities comparable with vacuum-coated replicas, and their contrast is enhanced by the dyes in the tape and liquid.

9.4.4 *Taper sectioning*

Features that occur close to the surface of a material, e.g. coatings, diffusion zones, fatigue damage, can be examined with greater ease by taper sectioning. This involves cutting through the surface of interest at an oblique angle, so that the depth of the surface layer and the component of the surface topography normal to the surface are geometrically magnified. Thus if the surface of the specimen is set at a taper angle α to the plane of polish (Fig. 9.4*a*) the depth is magnified by a factor cosec α.

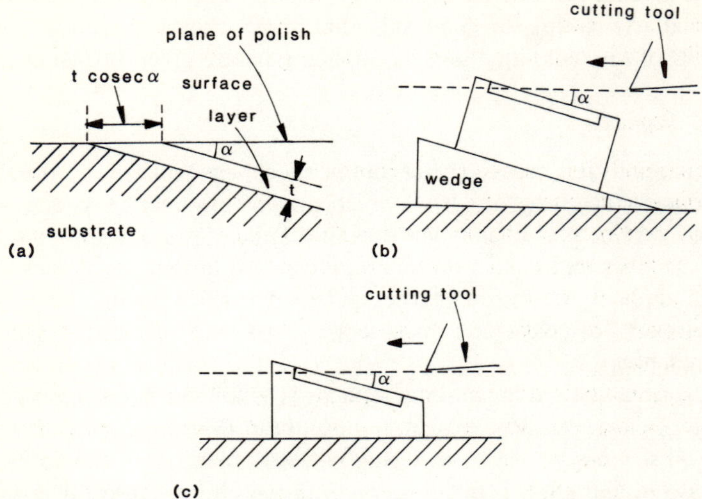

Figure 9.4 (*a*) Schematic diagram showing the principle of taper sectioning; (*b*) and (*c*), methods of obtaining a desired taper ratio. In (*b*) the specimen is set either on a wedge of the desired angle or held in a block with a slanted hole and cut, while in (*c*) the specimen is mounted in plastic above a wedge of the desired taper angle and cut.

Before taper sectioning the specimen needs to be heavily electroplated to protect the surface. Then a plane of polish is established at the desired small taper angle to the surface. This may be done either by setting a mounted specimen on a wedge of the desired angle or in a block with a slanted hole at the desired angle, and surface cutting or grinding the surface (Fig. 9.4*b*), or by using a wedge to set the taper angle during mounting (Fig. 9.4*c*), before cutting or grinding. Because it is difficult to set the taper angle accurately, the precise taper magnification is best found by mounting with the specimen a fine wire of 0.15–0.2 mm diameter perpendicular to the line of sectioning, and measuring the major axes of the elliptical section in the plane of polish. Often a taper magnification of $\times$ 10 is used, for which α is 5.74°.

10 Photomicrography

Photomicrography is the photographic recording of the image produced by a microscope; it should be clearly distinguished from microphotography, the photographic reproduction of the images of large objects on a fine scale. It lacks the flexibility of visual observation, thus we need to examine the specimen thoroughly before selecting a representative field for reproduction. Moreover, the information that we can obtain by subtle use of focusing is denied us. However, resolution of detail can sometimes be perceived more easily in a micrograph than curing visual observation.

10.1 Principles

Image formation in photomicrography is shown schematically in Fig. 10.1, and the magnification, M, produced is

$$M = \frac{h''}{h} = \frac{h'}{h} \times \frac{h''}{h'} = \frac{t}{f_0} \times \frac{c}{f_e} \tag{10.1}$$

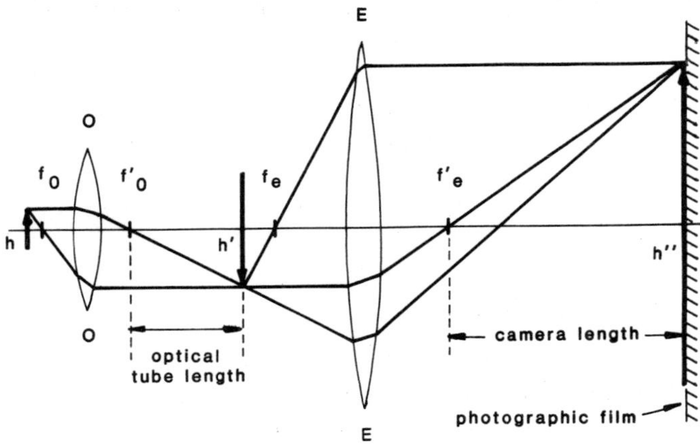

Figure 10.1 Formation of the projected image.

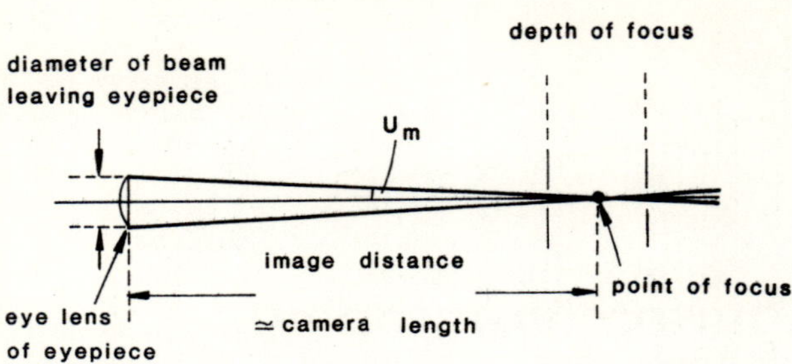

Figure 10.2 Criterion for the depth of focus of the projected image.

where h = object size, h' = primary image size, h'' = final image size, t = microscope optical tube length, c = camera length, f_0 = objective focal length, and f_e = eyepiece focal length.

Thus the magnification is proportional to the camera length, and it can be changed by altering the camera length. Theoretically, the camera length is the distance from the back focal plane of the eyepiece to the image plane, but because the focal plane is close to the eyelens it is acceptable to measure it from the eyelens. However, if we required a precise value for the magnification it is better to determine it by photographing a stage micrometer.

The depth of focus in the image space, D, is given by the equation

$$D = \frac{\lambda}{n' \sin^2 U_m} \tag{10.2}$$

where n' = refractive index of the medium in the image space, usually unity, and U_m = the angle the marginal rays make with the optic axis (Fig. 10.2).

In the case of the projected image U_m is very small. Typically the diameter of the beam leaving the eyepiece is 3 mm, thus if the image is formed 250 mm from the eyepiece $\sin U_m \simeq \tan U_m = 1.5/250 = 0.006$, and the depth of focus is ~ 15 mm for green light of wavelength 550 nm. Hence the location of the photographic film along the camera axis is not critical with respect to accurate focusing, but the magnification is affected if the film position is altered. Moreover, if the image is formed very close to the eyepiece the objective is made to work under conditions for which it was not corrected and the image is degraded.

10.2 The projection microscope

While it is possible to attach a camera to a bench microscope in order to produce photomicrographs, it is more convenient to use a projection microscope that is designed for the task. Modern projection microscopes developed from an early arrangement in which the illuminating system, the

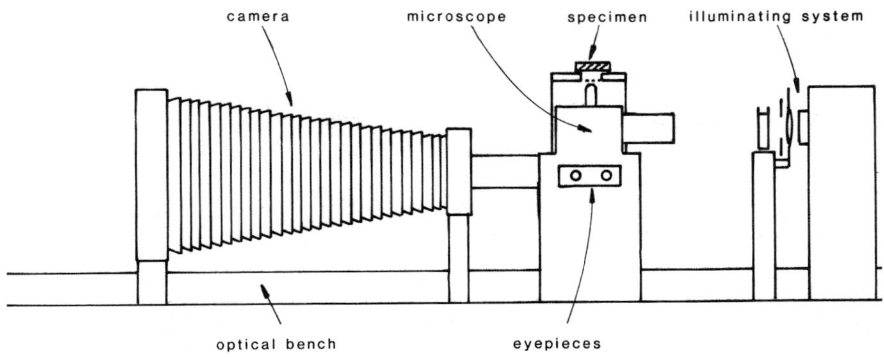

Figure 10.3 Diagram of a projection microscope mounted on an optical bench. Modern, more compact projection microscopes were developed from this type of instrument.

microscope and the camera system were on a well-supported optical bench (Fig. 10.3). The light from the specimen passed through the objective and could be either reflected into an eyepiece for visual examination, or onto the ground-glass screen of a plate camera. In modern projection microscopes the camera system is often incorporated in the body unit of the microscope, and may be of either fixed or variable length. Separate provision is often made for recording the image on either photographic plate or sheet film, or on 35 mm film. Moreover, this type of microscope is built to take special attachments so that specimens can be examined using a wide variety of special techniques such as interference contrast. Since the widespread adoption of 35 mm film for photomicrography, several microscope manufacturers have produced much heavier versions of the bench microscope that incorporate facilities for photomicrography using this type of film. They are considerably cheaper than the type of projection mircoscope described above, but often lack its versatility.

If it is necessary to use a bench microscope with a conventional 35-mm camera, preferably the camera should be supported independently of the microscope. The film plane must be perpendicular to the optic axis of the microscope, and the eyepoint of the microscope should be at the centre of the front surface of the camera lens. A matt black flexible tube should connect the microscope to the camera lens to exclude extraneous light. The camera lens is focused close to infinity, because the microscope is designed to work with its image at infinity. In use the microscope is focused visually, keeping the eye relaxed, and the camera then placed above the eyepiece without disturbing the microscope. It is advantageous, but not essential, to use a single lens reflex camera, because focus in the film plane can be observed directly. Commercial single lens reflex camera microscope adaptors that are simple tubes arranged to locate the film plane about 50 mm above the eyepiece without the camera lens in place are unsatisfactory, because the microscope has to be grossly maladjusted to produce the image in this plane, causing the lens corrections to be ineffective.

10.2.1 *Illumination, adjustment and focusing of the microscope*

Specimens should be uniformly illuminated by a steady, intense, compact light source. Mercury, xenon, and quartz-iodine lamps are widely used as light sources. They give out a considerable amount of heat and it is necessary to include in the incident light beam a heat-absorbing filter, either a cell filled with water or slightly acidified 1% copper sulphate solution, or heat-absorbing glass. Moreover, since uniform illumination is essential, Köhler illumination is prefered. The components of the illuminating system must be clean and the stops in it adjusted for optimum illumination as described in Chapter 4.

The focusing procedure depends on the size of the photographic negative to be produced. For large-format negatives, say 85 mm × 110 mm or larger, the image is focused on a ground-glass screen. However, because ground glass has a grainy texture it reduces the image brightness and resolution, making the focusing of fine structures difficult. These are best focused on the clear area with a central cross, which is often provided in the centre of the screen. The image is observed by using a magnifier *held at a fixed distance* that is focused on the cross. If no clear area on the screen is provided, one can be made by pencilling a cross at the centre of the screen on the ground surface and cementing a coverslip over it.

For miniature (35-mm) negatives the camera is equipped with a viewing screen incorporating a format reticule that is viewed through a focusing magnifier. The magnifier must first be sharply focused on the reticule, before focusing the image on the screen. Failure to focus on the reticule is an important cause of unsharp negatives. With low-power objectives the focus can be checked by testing for parallax between reticule and image; if the relative positions of the two remain unchanged the image is correctly focused.

In all cases the focus should be checked and, if necessary, adjusted finally with the filters to be used during photography in place.

10.2.2 *Vibration*

Vibration during exposure of the film causes unsharp negatives. Therefore, the microscope must be protected from vibration. Ideally the microscope should be set on a massive support and be located in a part of the building where vibrations from external sources are least. In severe cases of external vibration the microscope and its support should be mounted on damping material, such as foam rubber. Sometimes internal vibrations arising from the shutter mechanism may occur. These are minimized by using a Compur (iris) shutter in preference to a focal-plane shutter, because the blades in the Compur shutter move in a symmetrical manner. Moreover, the effects from this source of vibration are reduced by the use of longer exposures. Also when making an exposure it is expedient not to rest on the microscope support.

10.3 Exposures and light filters

As for conventional photography, the exposure required depends on the speed of the film used, the exposure time, and the intensity of the illumination. In turn the latter depends on the light source used, the NA of the objective, the eyepiece magnification, and the camera length. Moreover, if the exposure times are long reciprocity failure occurs, which makes it necessary to use longer exposure times than those deduced from the film speed.

10.3.1 *Determination of exposure time*

On microscopes not fitted with exposure meters, the correct exposure time is found by first making a series of trial exposures, developing and fixing the film, and selecting by inspection the exposure time that has yielded the negative with the best gradation. When using large-format film the procedure adopted to obtain the series of trial exposures on a single sheet of film is as follows. Having focused the image, the ground-glass screen is replaced by the film holder and, with the shutter closed, the dark slide is withdrawn. Successive exposures should differ by a factor of two, therefore suppose that the exposure series selected is 1, 2, 4, 8, and 16 s. The whole film is first given an exposure of 1 s, then the dark slide is reinserted about 20 mm into the holder. A second exposure of 1 s is made, the dark slide moved a further 20 mm and so on. Each successive exposure time is made equal to the sum of the previous exposures, so that successive strips of the film receive the selected exposures. When 35-mm film is used, exposing the film strip by strip is impractical and the several exposures must be made on separate frames of film.

When the microscope is fitted with an exposure meter especially designed for use in photomicrography, the exposure can be determined easily and accurately for each specimen. Some exposure meters need to be calibrated before use, using test film as described above. However, an increasing number of modern exposure meters are precalibrated and can be adjusted for use with films of different speeds. Nonetheless, when using a new kind of film it is wise to make additional exposures at half and at twice the indicated exposure, compare the negatives obtained, select the optimum exposure, and, if necessary, adjust the film speed rating.

10.3.2 *Assessment of optimum exposure*

The optimum exposure is determined by inspecting the lighter and darker areas of the negative, or positive if reversal film is being used. All the detail in these areas should be apparent. Underexposure causes loss of detail in the lighter areas, while overexposure causes lowering of contrast and loss of detail in the darker areas. With experience, assessment of optimum exposure is not difficult.

10.3.3 *Light filters*

The light used for photomicrography must often be modified in order to reveal the structure most clearly. This is achieved by introducing into the illuminating system filters that are usually made from coloured glass or gelatin. Two types of filter, selective and neutral density, are used.

Selective filters filters provide light of the desired range of wavelengths by absorbing the unwanted parts of the spectrum. An extensive selection of filters in a wide range of colours is cheaply available because they are widely used in photography.

When working with black and white film, yellow-green or green filters are most often used, because achromatic objectives are best corrected for, and the eye is most sensitive to, this part of the spectrum. Moreover, since the emulsions of photographic films, no matter of what type, have spectral responses that differ markedly from that of the eye, use of these filters brings the responses of the films closer to that of the eye, and results in images that match more closely those seen by the eye.

Colour filters also can be used to enhance or modify the contrast of the image. For example, in some aluminium bronzes, blue-grey particles of γ_2 phase are dispersed in a golden-yellow matrix of α phase, which appears lighter than the γ_2 phase to the eye. Photographed using a yellow-green or green filter they appear equally bright, but photographed using a deep orange filter they appear with a contrast similar to that seen by the eye.

Moreover, using apochromats or planapochromats modest enhancement of the resolution can be achieved by using a blue filter. However, use of blue light makes focusing more difficult, because the eye is less sensitive to it, the intensity of blue light in the illuminating beam may be low, and the exposure may be appreciably altered.

When working with colour film, use of colour-compensating filters is essential. They ensure that the colour temperature of the illumination corresponds to that for which the film is balanced, and they can be used to adjust the colour balance of the film.

Neutral density filters are used to reduce the intensity of the illumination uniformly across the spectrum in cases where the voltage across the light source cannot or may not be altered. Thus in colour photomicrography the bulb must be operated at a specific voltage in order for it to emit light of the desired colour temperature.

10.4 Black and white photography

10.4.1 *Formation of the photographic image*

A photographic film or plate consists of a layer of light-sensitive 'emulsion' supported on a substrate of plastic film or glass plate. The 'emulsion', which is not a true emulsion, is a dispersion of microscopic silver bromide-iodide grains (crystals) in gelatin. When the emulsion is exposed to light, grains of silver halide that receive more than a certain minimum exposure are sensitized and form a latent image. During development the sensitized grains are reduced to metallic silver by the mild reducing agents in the photographic developer solution, yielding a photographic negative. The areas that have received the greatest exposure have the greatest optical density and vice versa, thus the image formed is complementary to that of the object. Not all the silver halide is reduced by the developer and that which remains must be dissolved by the fixing solution. Subsequently the emulsion is washed to remove the fixing agent, and dried, thus giving a permanent negative of the image.

10.4.2 *Emulsion characteristics*

The characteristic curve and contrast. The exposure received by a film, E, is the product of the intensity of the illumination, I, and the exposure time, t, i.e.

$$E = I \times t \tag{10.3}$$

This equation is called the reciprocity law; it implies that the same exposure is obtained whether we expose the film to intense illumination for a short time or to weak illumination for a long time. It is obeyed over a fairly wide range of intensities and times, but it breaks down when the illumination is either very intense or very weak. This is called reciprocity failure, and under these conditions often very much greater exposure is needed to achieve the level of exposure predicted by the law. In practice, the law is obeyed for

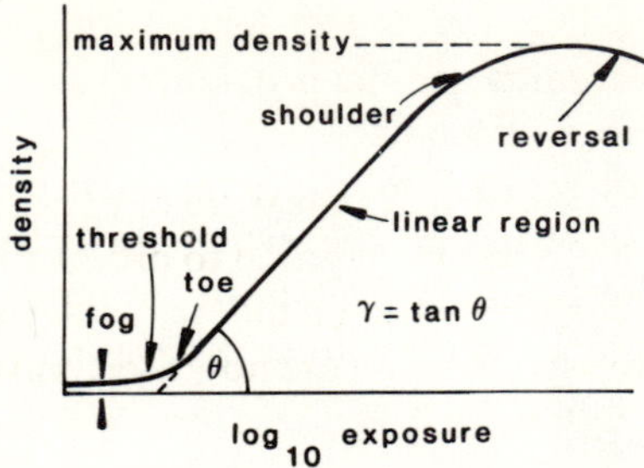

Figure 10.4 Schematic diagram of the characteristic curve of a photographic emulsion.

exposure times in the approximate range 1/1000 to 1s for most emulsions.

The relationship between the optical density of the negative and exposure is represented by the characteristic curve of the film (Fig. 10.4). The main parts of the curve are the threshold, the toe, the linear region, the shoulder, and the region of reversal or solarization. The most important region is the linear one, and its extent and slope depend on the kind of film, the developer and the developing conditions used. Its slope, $\gamma = \tan\theta$, is a measure of the contrast of the negative. In exposing a film we aim to have the range of exposures lying in the linear region. If some exposures fall on the toe of the curve the film is underexposed, and if on the shoulder over-exposed.

Colour sensitivity. Early photographic emulsions were only sensitive to blue and near ultra-violet light, and often gave incorrect rendering of tones because they were insensitive to green and red light. However, by the use of suitable dyes emulsions are made sensitive to green or green and red light. Emulsions that are sensitive to blue and green light are described as orthochromatic, while those that are sensitive to blue, green and red light are described as panchromatic. The films and plates used for photo-micrography are either orthochromatic or panchromatic, nowadays usually the latter.

Resolving power. Photographic emulsions need to have high resolving power in order to show the detail in the resolved image. The resolving power of an emulsion depends upon the size and distribution of the clumps of silver grains in the emulsion, i.e. the granularity, rather than on the size of the individual grains, which is very small, e.g. about 30 nm in high resolution film. For photomicrography the granularity of the emulsion needs to be fine or very fine. On the basis that the smallest distance between two resolved points in the image should be approximately ten times the granularity of the film, taken as 10 μm, Martin and Johnson (30) recommend that the *minimum* magnification on the negative should be 300 × NA.

Film speed. The speed of a film is a measure of its sensitivity to light; the faster the film the less the exposure required to produce a correctly exposed negative. A number of systems based on different criteria have been used to define film speed, but the systems that are now used universally are the ASA or BS system, and the DIN system. Both systems are based on the exposure needed to give a fixed density above fog level. The ASA system uses arithmetical units, thus a doubling of the speed number indicates a doubling of the speed, while the DIN system uses logarithmic units, in which an increase of three units in the speed number indicates a doubling of the speed.

Exposure latitude. Exposure latitude is defined as the factor by which the minimum exposure required to give a negative showing adequate detail in the least dense areas of the negative can be multiplied without losing detail in the most dense areas. The greater the latitude the wider the range of exposures that will give a satisfactorily exposed negative.

10.4.3 *Selection of films for photomicrography*

Films should be fine-grained and of medium or high contrast, depending on the type of image to be recorded. The speed of the film is not usually important, since the image is stationary. In practice, high-speed films are relatively coarse-grained, thus usually medium- or low-speed films intended for conventional photography are used, e.g. except for Fig. 6.3 the photomicrographs in the book were taken on 35 mm Ilford FP4 film, that has a daylight speed of 125 ASA, medium contrast and is very fine-grained, and the film was developed in ID 11 developer. For most purposes, there is little to choose between the quality of photomicrographs recorded on large-format film and 35-mm film, provided that the images have been correctly focused and the films correctly exposed and carefully processed; the former are more tolerant of imprecise focus because they usually do not need to be enlarged.

As an alternative to conventional films, self-developing materials may be used in large-format cameras to produce instant-prints.

10.4.4 *Photographic papers*

Photographic papers consist of a paper base coated with a fine-grained chloro-bromide or bromide emulsion that is insensitive to the red end of the spectrum, so that they can be processed in low-intensity orange or yellow light. Most papers are available in a number of contrast grades, commonly five; grade 1 (soft) paper has the least contrast and grade 5 (extremely hard) paper the greatest. Most properly exposed and developed negatives print satisfactorily on grade 2 (normal) paper. There are also variable-contrast, 'multigrade', papers whose contrast is controlled by the the colour of the printing light. The base may be of either single or double weight, and a variety of surface finishes is available, but a glossy finish is normally used for photomicrographs. Ordinary glossy papers need to be glazed to develop their greatest gloss. A fairly recent innovation is resin-coated papers, which have a number of advantages over ordinary papers including much shorter washing and drying times, and elimination of the need to glaze glossy papers. However, they are more easily damaged in processing when several sheets are processed simultaneously, and the surfaces of the prints may craze if exposed to light over a long period of time.

10.4.5 *Development of films*

Films are best processed in light-tight tanks, because most modern materials are panchromatic, and must be processed in total darkness. They should be developed for a specified time at a particular temperature, i.e. using the time-temperature method. The development time needed to produce correct development is determined mainly by the film used, the dilution, state and temperature of the developer chosen, and the amount of agitation using during development.

After loading the film into the tank in total darkness, processing may be carried out in full light. The film is developed in the chosen developer at constant temperature, usually 20° C, for the time recommended with gentle agitation at regular intervals throughout the development period, e.g. 5 s in every minute. At the end of the development period the developer is poured off and the film rinsed with water *at the same temperature** to remove residual developer solution or, better, with an acid stop bath. Then it is fixed using an acid fixing bath *at the same temperature** which is gently agitated. After fixing for the recommended time, 10–20 minutes for ordinary fixer, or 1–2 minutes for rapid fixer, the fixing solution is poured off and the film is washed in water *at the same temperature** to remove fixing salts. Washing may be done in running water for about 30 minutes, or by using six changes of water of five minutes each without agitation, or six changes of water of two minutes each with agitation. Drying marks on the film are minimized

* It is important that the film should not be subjected to rapid temperature changes, which may cause reticulation, i.e. a fine irregular wrinkling of the emulsion.

by adding a few drops of wetting agent to the final rinse, and removing excess water from the film with a soft sponge or squeegee. After the final rinse and removal of excess water from the film, it is dried by hanging in a dust-free place at room temperature. Rapid drying by heating is not recommended, unless a specially designed drying cabinet is used.

The developer is essentially an alkaline solution of mild reducing agents, e.g. hydroquinone, metol, that reduce the sensitized silver halide grains to metallic silver in a controlled manner. By choosing a developer containing appropriate reducing agents the contrast and the granularity of the negative can be controlled within limits. Control of the temperature at which a developer is used is important because altering it by 1°C changes the development time by about 10%. Moreover, films should not be developed outside the temperature range 13°–24°C, because below 13°C development is very slow, while above 24°C the film emulsion may soften and be damaged. The stop bath is a dilute solution of a weak acid, e.g. acetic acid, that neutralizes the developer solution and stops development completely. The fixing bath is a sodium or ammonium thiosulphate solution that dissolves the unreduced silver halide in the film. Usually, the bath also contains a weak acid that neutralizes residual developer in the emulsion and hardens the emulsion.

10.4.6 *Printing of negatives*

Processing is carried out in the light of an orange or yellow safelight. Positive prints are prepared by either enlarging or, for large-format negatives, contact printing. In enlarging an image of the negative is projected onto the photographic paper, while in contact printing the paper is exposed to light through the negative, whose emulsion side is placed in contact with the emulsion side of the paper. In both cases the correct exposure is determined by trial, in the same way as described for negatives. The exposed paper is developed in a print developer, e.g. D163 or ID20, with agitation, fixed, washed, usually in open dishes, and dried. Development times, usually 1½ to 3 minutes, are much shorter than those used for films.

The quality of the prints produced depends on both the films and prints being correctly processed. In particular, prints should always be developed for the recommended time; overexposed prints should not be saved by stopping development prematurely, because the loss in quality of the print is always clearly apparent.

When negatives are enlarged the magnification produced by the enlarger must be taken into account when stating the magnification of the micrograph.

10.5 Colour photography

Modern colour films, both negative and reversal, are made up of three layers of photographic emulsion that are sensitive to blue, green and red light, and are separated by appropriate filters. During development, images corresponding to the three colours are formed in the layers and are converted to appropriate dye images by complex chemical reactions. The processing of the film is much more complex than that of black and white film, but in some cases can be done in the laboratory.

Both amateur and professional colour films are suitable for photomicrography. The former usually must be returned to the manufacturer for processing. Details of the processing procedures for films that may be developed in the laboratory depend on the kind of film used and the developing system that must be used with it. Whatever system is used it must be followed with great attention to detail. Self-developing colour materials that give instant prints are also available and may be used for photomicrography with large-format cameras. However, owing to the relatively high costs of colour materials, colour photomicrographs usually are made on 35-mm film.

Fine-grained colour films should be used and these are of low or medium speed. Particular care must be taken to illuminate the specimen with light of the colour temperature for which the film is balanced. However, films from different manufacturers and even different films from the same manufacturer do not reproduce the colours in exactly the same way. Consequently we may need to redress the colour balance by use of colour correction filters.

References and further reading

1. L.C. Martin, *The Theory of the Microscope*, London and Glasgow, Blackie, 1966.
2. B.O. Payne, *Microscope Design and Construction*, York, Vickers Instruments Ltd., 1956.
3. W.G. Hartley, *Hartley's Microscopy*, Charlebury, Senecio, 1979.
4. J.H. Richardson, *Optical Microscopy for the Materials Sciences*, New York, Dekker, 1971.
5. N.H. Hartshorne and A. Stuart, *Practical Optical Crystallography*, London, Arnold, 1969.
6. N.H. Hartshorne and A. Stuart, *Crystals and the Polarizing Microscope*, London, Arnold, 1970.
7. G.K.T. Conn and F. Bradshaw, *Polarized Light in Metallography*, London, Butterworths, 1952.
8. Anon., *Metals Handbook*, 8th edn., Vol. 8, Metals Park, Ohio, American Society for Metals, 1973.
9. G.L. Kehl, *Principles of Metallographic Laboratory Practice*, New York, McGraw-Hill, 1949.
10. R.C. Gifkins, *Optical Microscopy of Metals*, London, Pitman, 1970.
11. H. and S. Modin, *Metallurgical Microscopy*, London, Butterworths, 1973.
12. S. Tolansky, *Multiple Beam Interference Microscopy of Metals*, London and New York, Academic Press, 1970.
13. C.E. Price and D. Cox *Metal Progress*, 1983, **123**, (2), 37.
14. H. Insley and V.D. Frechette, *Microscopy of Ceramics and Cements*, London and New York, Academic Press, 1955.
15. H.G. Smith and M.K. Wells, *Minerals and the Microscope*, London, Murby, 1973.
16. C.H. Leith-Dugmore, *Microscopy of Rubber*, Cambridge, Heffer, 1961.
17. D. Hemsley, *The Light Microscopy of Synthetic Polymers*, Oxford, Oxford University Press, 1984.
18. E.E. Underwood, *Quantitative Stereology*, Reading, Mass., Addison-Wesley, 1970.
19. F.B. Pickering, *The Basis of Quantitative Metallography*, London, The Institute of Metallurgists, 1976.

20. T. Gladman and J.H. Woodhead, *J. Iron and Steel Inst.*, 1960, **194**, 189.
21. T. Gladman, *J. Iron and Steel Inst.*, 1963, **201**, 1044.
22. L.E. Samuels, *Metallagraphic Polishing by Mechanical Methods*, Metals Park, Ohio, American Society for Metals, 1982.
23. P.A. Jacquet, 'Electrolytic and chemical polishing', *Metallurgical Reviews*, 1956, **1**, 157.
24. W.J.McG. Tegart, *The Electrolytic and Chemical Polishing of Metals in Research and Industry*, Oxford, Pergamon Press, 1966.
25. M. Beckert and H. Klemm, *Handbuch der Metallographischen Ätzverfahren*, Leipzig, VEB Deutsches Verlag für Grundstoffindustrie, 1962.
26. G. Petzow, *Metallographic Etching*, Metals Park, Ohio, American Society for Metals, 1978.
27. C.J. Smithells, *Metals Reference Book*, 6th edn., London, Butterworths, 1982.
28. T.R. Allmond, *Microscopic Identification of Inclusions in Steel*, London, British Iron and Steel Research Association, 1962.
29. H-E. Bühler and H.P. Hougardy, *Atlas of Interference Layer Metallography*, Oberursel, Deutsches Gesellschaft für Metallkunde, 1980.
30. L.C. Martin and B.K. Johnson, *Practical Microscopy*, London and Glasgow, Blackie, 1958.
31. R.E. Jacobson *et al.*, *The Manual of Photography*, London, Focal Press, 1978.

Appendix: The care of the microscope

Microscopes are delicate instruments and care should be taken to avoid knocking them and, especially, to avoid dropping eyepieces and objectives, which are easily damaged. They should be kept free from dust and grease, and should be covered when not in use. The top of the microscope tube should always be closed, either with an eyepiece or a rubber plug to exclude dust. Similarly, any unoccupied positions on the objective changer should be closed by blanking plugs. The focusing mechanisms should be kept clean and free from coagulated grease. They should not be forced: before beginning to focus it is helpful to ensure that the fine focus is near the centre of its range of travel.

Optical components shoud be kept free of dust, grease and finger marks. Dust may be removed by the use of a rubber blower or blower brush. However, brushes quickly become contaminated with grease and may then spread grease onto the surfaces being cleaned, thus frequent washing of brushes used for this purpose is desirable. Removal of dust from a lens surface also may be achieved by the use of a brush warmed on an electric light bulb, which will attract dust particles from the surface. Dust particles lying in or near conjugate planes in the optical system that are brought to focus in the image plane are objectionable because they appear in the image as unsharp discs or worms and are very noticeable in photomicrographs. When such artefacts occur, the site of the dust should be identified and the dust removed, e.g. dust in the eyepiece can be identified by rotating the eyepiece and observing if the artefacts also rotate. Dust particles elsewhere in the optical system may contribute to glare.

Grease and finger marks may be removed by carefully wiping the lens surfaces with lens cleaning tissue moistened with one or two drops of xylene*, or acetone. Nevertheless, lenses should not be washed in these

*Xylene is toxic.

solvents because the cement holding the lenses may be attacked. Alcohol should never be used for cleaning lenses because its attacks the cement very readily. Grease on the eye lens of an eyepiece accumulates rapidly, owing to eyelashes brushing the lens, and frequent cleaning is necessary. Finger marks can be avoided by adopting careful working procedures. It is permissible to unscrew eyepieces for cleaning or insertion of graticules, but objectives and other lenses should never be taken apart; only the front and back surfaces of such lenses should be cleaned.

Damage to both objective and specimen may be caused by the objective being racked into the specimen. This can be avoided by bringing the objective closer to the specimen than the point of focus and racking it away from the specimen until focus is achieved. For this purpose it is helpful to remember which way the focusing knobs must be turned to move the microscope away from the specimen. Moreover, with a microscope with parfocal objectives the lowest power objective always should be focused first because it is the easiest to focus. Higher power objectives can then be brought beneath the microscope tube and require only slight adjustment of the fine focus to achieve sharp focus.

When immersion objectives are used one small drop of immersion oil should be applied to the front lens of the objective and one to the specimen. The objective and specimen are then brought together until the drops coalesce and focusing is completed using the fine focus mechanism. After use the objective should be cleaned carefully with lens tissue to remove the immersion oil. With modern, non-drying immersion oils it is possible to remove the oil with lens tissue without resort to the use of solvents. However, if the immersion oil has caked, as occurred readily with traditional cedar-wood immersion oil, it can be removed by careful repeated cleaning with lens tissue moistened with solvent.

INDEX